✓ clw.

QS-9000

ANSWER BOOK

101 Questions and Answers About the Automotive Quality System Standard

ROB KANTNER

John Wiley & Sons, Inc.

New York • Chichester • Brisbane • Toronto • Singapore • Weinheim

Library of Congress Cataloging-in-Publication Data:

Kantner, Rob.
   QS-9000 answer book : 101 questions and answers about the automotive
quality system standard / Rob Kantner.
      p.   cm.
   Includes index.
   ISBN 0-471-15700-7 (cloth : alk. paper)
   1. QS-9000 (Standard)—Miscellanea.   2. Automobile industry and
trade—United States—Quality control—Standards.   3. Automobiles—
Parts—Design and construction—Quality control.   I. Title.
TL278.K36   1996
629.23'4'0685—dc20   96-17931

Acknowledgments

Most of what I know about QS-9000 and ISO 9000 I learned not from books or classes, but on shop floors and in the offices of the many fine companies run by my clients, present and past. I am grateful to the following people, as well as all the others who have taught me so much.

Michael Bank	Tony Godfrey	Steve Smithson
Megan Betley	Bruce Harron	Paul Spina
John Cadarian	Jan Hemme	Tim Spina
Paul Castorina	Scott Horne	Jack Waters
Ken Cohen	Terry Leach	Clarence Watts
John Combites	Bob Papp	Jay Zimmern
Shirley Daft	Clyde Pearch	Jerry Zoli
John Doran	Marc Prigozen	

I have also had the unstinting help of many knowledgeable ISO 9000 and QS-9000 assessors, authors, and consultants. They know who they are, and how grateful I am to them.

R. K.

Contents

IMPROVING THE PROCESS 179

POST-PROCESS FUNCTIONS 195

PART FIVE:
IMPLEMENTING A
QS-9000 SYSTEM 203

PART SIX:
REGISTRATION 253

PART SEVEN:
AFTER REGISTRATION
279

APPENDIXES

INDEX
297

Introduction

Among suppliers to the Big 3 auto companies, there is a tradition of . . . how shall I put this . . . skepticism.

As the Big 3 and other major OEMs strive to cut costs and improve quality and efficiency, they have steadily increased pressure of all kinds on their suppliers. Each new initiative from the automotive giants generates responses that tend to be, to say the least, guarded. "What are they up to now?" some suppliers wonder.

QS-9000, the "new," "unified" automotive quality system standard, is just such an initiative. Like most business initiatives, it has caused some confusion, anxiety, and no small amount of controversy.

But there is a method to the QS-9000 "madness." That method is the underlying structure of the ISO 9000 quality system process. Suppliers may feel they are being unduly pressured by the Big 3, and it is not my place to comment on that. But what I do know is that, when it comes to QS-9000, the Big 3 are on the right track.

QS-9000 can help automotive suppliers improve their performance—if they implement it properly.

QS-9000 is partially made up of automotive-related quality methods and sub-systems that have been in use for years. And, it's partially drawn from the international ISO 9001 standard, first published in 1987 (itself, the latest generation in a long evolution of quality system standards). QS-9000 is already the requirement for third-party audits and registration for a large portion of the supply chain.

Nothing new individually. But taken together, QS-9000 is very new to the industry. That is the cause of the hesitation, reluctance, and skepticism. "Who are they to tell us what to do?" many suppliers cry. "Do they have the right to tell us how to run our business?"

Well, yes, they do. But only if you acknowledge that the "they" here are your customers—and if you do not keep your customers satisfied, they will go elsewhere.

For many suppliers who want to keep their customers, QS-9000 is the new "rule of the road." Deadlines have been drawn in the sand, and for many suppliers, time is growing short. They are a-bubble with questions.

- What *is* QS-9000?
- Does it apply to us?
- How do we implement it?
- What is "registration" all about?
- What can QS-9000 do for us?

The *QS-9000 Answer Book* is an effort to answer these questions. It is meant to be a quick, easy to digest, guide to the basics of the QS-9000 process. It is aimed at management and supervisory people who may, up until now, have had little if any exposure to the "quality" disciplines. For that reason, I have tried to avoid using technical jargon and "quality-speak."

Using my years of experience implementing QS-9000 and ISO 9000 quality systems for client companies all over the country, I have compiled a guide that is handy and practical. Much of the book focuses on achieving "registration" to QS-9000, since that is the goal of many companies today. But I see "registration" as an interim goal. I believe that a company's ultimate goal is to design and manage a system that continuously improves the performance of the company. A QS-9000 system can do that, if you let it. This book is intended to show you how.

Please keep in mind that QS-9000 is very new. It was first published in August, 1994, and was slightly revised in February, 1995. As this book goes to press, only a relative handful (fewer than 500) companies have been all the way through the implementation and registration process. For this reason, it is still unclear as to how many of the requirements need to be interpreted and implemented in response to specific business needs.

This is my way of cautioning you that you must not regard this book as the ultimate authority on QS-9000. Your company and situation are unique. Address your specific technical questions to your registration body or qualified consultant.

How to Use This Book

The 101 questions and answers in this book address the essential issues related to QS-9000: the requirements, registration, implementation, and more.

Requirements are clearly identified as such. These must be adhered to without exception. "Guidance" is clearly identified as such. Much of the guidance information is drawn from my own experience in helping companies implement QS-9000 and ISO 9000 systems. Other guidance is drawn from published ISO 9000 guidance documents, which are just as relevant to QS-9000.

Each question has a "capsule answer." This provides just a kernel of information and is meant to be a browsing aid and a memory jogger. In addition to explanations, technical requirements, and technical guidelines, many answers also include audit-related information, as well as details on documents that are required.

Although the questions and answers follow a logical sequence, it is not necessary to read the book that way. Each answer stands more or less on its own. Abundant cross-references let you flip from answer to answer so you can access the information you need in the way that suits you best.

Finally, the knowledge and experience in this book is updated every day, thanks to the kind help of clients, colleagues, and readers. If you have any suggestions or questions, feel free to contact me via the publisher or via my home page on the World Wide Web: http://www.cris.com /~rob4334/iso.htm.

As is the case with my earlier book, *The ISO 9000 Answer Book*, the *QS-9000 Answer Book* gives you fast, easy access to most of the information you need to know.

ROB KANTNER
Wayne, Michigan

CROSS-REFERENCE CHART:
QS-9000 REQUIREMENTS AND THE RELEVANT QUESTIONS

Element	Title	Question
4.1	Management Responsibility	22, 23, 24, 28, 31, 68, 75
4.2	Quality System	17, 30, 34, 35, 36, 49, 55, 74
4.3	Contract Review	40
4.4	Design Control	45, 46, 47, 48, 50
4.5	Document and Data Control	34, 38, 49
4.6	Purchasing	43, 44, 56
4.7	Control of Customer-Supplied Product	41
4.8	Product Identification and Traceability	53
4.9	Process Control	37, 49, 54, 56, 57, 58, 59, 60
4.10	Inspection and Testing	61
4.11	Control of Inspection, Measuring, and Test Equipment	62, 63, 64
4.12	Inspection and Test Status	61
4.13	Control of Nonconforming Product	65
4.14	Corrective and Preventive Action	66, 67
4.15	Handling, Storage, Packaging, Preservation, and Delivery	69, 70
4.16	Control of Quality Records	33
4.17	Internal Quality Audits	32, 87, 88
4.18	Training	28, 29, 78
4.19	Servicing	41, 42
4.20	Statistical Techniques	68
II.1	Production Part Approval Process	39
II.2	Continuous Improvement	27
II.3	Manufacturing Capabilities	38, 51, 52

QS-9000 Overview

I. What is QS-9000?

QS-9000 is a written set of rules (a "standard") published by major American car and truck companies as a joint venture. The rules state the practices that the sponsors—mainly, the Big 3 auto companies: Chrysler Corporation ("Chrysler"), Ford Motor Company ("Ford"), and General Motors Corporation ("GM")—expect their main suppliers to implement. The sponsors buy billions of dollars' worth of products and services from outside suppliers each year. Their customers' needs will be met more consistently if all their suppliers follow the same set of rules.

CAPSULE ANSWER
QS-9000 is a set of "quality standard" requirements published by major car and truck manufacturers. By adhering to QS-9000, their direct suppliers of certain products and services become part of a commonly understood, independently verifiable quality management system.

The Big 3 auto companies go beyond expecting their main suppliers to implement these practices; they *require* it. Any direct supplier of production materials, production or service parts, or certain specified services to a Big 3 company must implement systems that are consistent with QS-9000 rules (referred to as "requirements"). In most instances, prior registration is necessary (Question 5).

Because "meeting customer needs" is one of many definitions of quality performance, QS-9000 is often called a "quality system" or a "quality management system." But the complete set of rules goes beyond quality matters as they are generally understood. Very generally, the requirements can be categorized as:

- Requirements that help ensure that the supplier's output meets the auto company's specifications and announced product descriptions (popularly referred to as quality control).
- Requirements that ensure that the supplier's quality system is consistently implemented and verifiable. Has the supplier done exactly what was agreed to and would its delivered product or service stand up under an independent, objective audit?
- Requirements that support the continuous improvement of the supplier's ability to meet the auto company's needs. Competing companies cannot stand still; they must constantly strive to get better results from better methods.

Nothing in QS-9000 is new. Roughly 60 percent of the requirements are drawn, word for word, from the relevant International Organization for Standardization quality standard, issued in 1994: ISO 9001 (Question 2). The remaining 40 percent refer to quality-related methods and practices that have been in use in the automotive industry for years. QS-9000 is intended to *simplify* the rules for suppliers—to harmonize and consolidate the sometimes conflicting, sometimes redundant "quality programs" each of the sponsors has imposed on its suppliers. The Big 3 quality programs are:

- Chrysler: Pentastar.
- Ford: Qi.
- GM: Targets for Excellence.

QS-9000 is not a wholly successful harmonization. But it is a major improvement, and that alone makes it a radical and remarkable departure for the automotive industry and a source of consternation among the industry's suppliers.

The mandate on adhering to QS-9000 represents a fundamental change in the way the Big 3 relate to their suppliers. As customers, the Big 3 have always wanted verification (proof) that their suppliers were following their prescribed quality rules. In the past, they verified their prescriptions directly, usually with site visits, reports, and audits.

QS-9000 establishes a new system for verifying that suppliers are meeting the requirements. Independent, third-party registration firms (Question 21) conduct audits of suppliers, confirm compliance to the QS-9000 standard, and register the suppliers. And the process does not stop there. To stay registered, suppliers must undergo semiannual surveillance audits, also carried out by their registration body.

By the end of 1997, registration will be required for all companies that supply certain products and services directly to Chrysler and GM (Question 5). Ford suppliers are not required to register by that cutoff date, but they are expected to comply with the QS-9000 requirements.

This shift of the burden of verification has shifted something else, too: cost. Under the old system, the auto companies funded the audit activity. Now, the suppliers do—and it is not cheap (Question 8).

The exact cost and level of difficulty for a supplier will depend on:

- The level of commitment of the supplier's senior management—the single most important factor.
- The ongoing condition of the supplier. A company that has disciplined, documented, well-implemented quality systems that include standard automotive quality practices will have at least a less difficult walk toward registration and use of QS-9000.
- Whether the supplier (or any part of it) has "design responsibility" (Question 45).
- The time frame. For a supplier that is under the customer's gun and has merely months to get a job done, the process will be highly stressful.
- The supplier's physical size and configuration.

When does QS-9000 apply to a supplier?

1. If it is a "Tier 1" supplier delivering the following directly to the Big 3:
   - Production or service parts.
   - Production materials.
   - Heat treating, painting, plating, or other finishing services.
2. If it is a significant supplier to a Tier 1 company. As a customer, the Tier 1 company may require implementation of QS-9000. The second-level ("Tier 2") supplier may be asked nicely, or receive a suggestion, or get no notice at all. Question 12 gives advice on being a supplier to a supplier.

The bottom line is this: QS-9000 is a very thorough, very prescriptive, somewhat unwieldy, sometimes annoying set of rules that several very major customers have posted for their suppliers to follow. A supplier may find the rules too expensive, exhausting, and irritating, and may walk away from those customers. However, a supplier willing to invest the time and effort to learn and to follow QS-9000 may improve its own internal standards of quality and land one or all of the Big 3 as ongoing customers.

2. What is the difference between QS-9000 and ISO 9000?

> ### CAPSULE ANSWER
>
> QS-9000 is in fact ISO 9000 plus a large list of additional and more prescriptive requirements.

The major difference is that ISO 9000 is meant to apply to any organization, making any product or service, anywhere in the world. QS-9000 makes ISO 9000 specific to the relationship between the major American-based car and truck companies and the international community of direct suppliers of specific parts and services to those companies.

QS-9000, then, is ISO 9000 *plus*—a hybrid of ISO 9001 requirements plus a lightly harmonized rehash of industry-specific requirements drawn from various Big 3 quality programs.

Part I of QS-9000 is, like ISO 9001, divided into 20 sections (or "elements"). Each carries the same name as the corresponding ISO 9001 element and uses the same precise wording. In the *Quality System Requirements QS-9000* manual, the text drawn from ISO 9000 is printed in italics.

Because all of the elements of ISO 9001 are included in QS-9000, registration to QS-9000 (Question 4) automatically includes registration to either ISO 9001 (if the supplier is responsible for design) or ISO 9002 (if it is not).

But each section of Part I of QS-9000 has additional specific requirements. These are minor in some cases (e.g., Element 4.3, Contract Review, has only one QS-9000 addition) and major in others (e.g., Elements 4.4, Design Control, and 4.9, Process Control). The additions are never inconsistent with the ISO 9000 language. Instead, they tend to tighten the requirements and make them more prescriptive.

The tabulation on pages 5–7 of Elements 4.1 through 4.20 of QS-9000 compares the ISO 9000 sections with the additions imposed by the automotive industry.

QS-9000 contains the following material not found in ISO 9000:

■ Part 2, "Sector Specific Requirements":
  – Production Part Approval Process.
  – Continuous Improvement.
  – Manufacturing Capabilities.
■ Part 3, "Customer-Specific Requirements":
  – Chrysler-specific requirements.
  – Ford-specific requirements.

Element	Title	ISO 9000 Sections	Automotive Additions
4.1	Management Responsibility	Quality Policy Responsibility and Authority Resources Management Representative Management Review	Organizational Interfaces Business Plan Analysis and Use of Company-Level Data Customer Satisfaction
4.2	Quality System	Quality System Procedures Quality Planning	Special Characteristics Use of Cross-Functional Teams Feasibility Reviews Process Failure Mode and Effects Analysis Control Plan
4.3	Contract Review	Review Amendment to a Contract Records	
4.4	Design Control	Design and Development Planning Organizational and Technical Interfaces Design Input Design Output Design Review Design Verification Design Validation Design Changes	Required Skills (plus supplementary requirements to ISO 9000 sections)
4.5	Document and Data Control	General Document and Data Approval and Issue Document and Data Changes	Reference Documents Document Identification for Special Characteristics Engineering Specifications
4.6	Purchasing	General Evaluation of Subcontractors Purchasing Data Verification of Purchased Product	Approved Materials for Ongoing Production Subcontractor Development Scheduling Subcontractors

(Continued)

5

(Continued)

Element	Title	ISO 9000 Sections	Automotive Additions
4.7	Control of Customer-Supplied Product	General	One additional "note."
4.8	Product Identification and Traceability	General	One additional "note."
4.9	Process Control	General	Government Safety and Environmental Regulations Designation of Special Characteristics Preventive Maintenance Process Monitoring and Operator Instructions Preliminary Process Capability Requirements Ongoing Process Performance Requirements Modified Preliminary or Ongoing Capability Requirements Verification of Job Setups Process Changes Appearance Items
4.10	Inspection and Testing	General Receiving Inspection and Testing In-Process Inspection and Testing Final Inspection and Testing Inspection and Test Records	Acceptance Criteria Accredited Laboratories Incoming Product Quality Layout Inspection and Functional Testing
4.11	Control of Inspection, Measuring, and Test Equipment	General Control Procedure	Inspection, Measuring, and Test Equipment Records Measurement System Analysis

Element	Title	ISO 9000 Sections	Automotive Additions
4.12	Inspection and Test Status	General	Product Location Supplemental Verification
4.13	Control of Non-Conforming Product	General Review and Disposition of Nonconforming Product	Suspect Product Control of Reworked Product Engineering Approved Product Authorization
4.14	Corrective and Preventive Action	General Corrective Action Preventive Action	Problem Solving Methods Returned Product Test/Analysis
4.15	Handling, Storage, Packaging, Preservation, and Delivery	General Handling Storage Packaging Preservation Delivery	Inventory Customer Packaging Standards Labeling Supplier Delivery Performance Monitoring Production Scheduling Shipment Notification System
4.16	Control of Quality Records	General	Record Retention Superseded Parts
4.17	Internal Quality Audits	General	Inclusion of Working Environment
4.18	Training	General	Training as a Strategic Issue
4.19	Servicing	General	Feedback of Information from Service
4.20	Statistical Techniques	Identification of Need Procedures	Selection of Statistical Tools Knowledge of Basic Statistical Concepts

   – General Motors-specific requirements.
   – Truck Manufacturers-specific requirements.

3. How is QS-9000 used?

The QS-9000 requirements and associated documents spell out a system. This system is intended to be the basis for the relationship, in terms of quality methods, between the Big 3 (and other) customers and their main suppliers (the Tier 1 suppliers).

> **CAPSULE ANSWER**
>
> QS-9000 is most often used by suppliers to meet the requirements of their Big 3 customers. But QS-9000 (or, if appropriate, ISO 9000) can bring benefits to virtually any company.

The QS-9000 system can also become the basis for the relationship between Tier 1 suppliers and *their* main suppliers. The Tier 1 companies have some discretionary leeway, but they are supposed to "perform subcontractor development" to the QS-9000 standard (Question 43).

QS-9000 is not meant to replace customer-specific requirements. Wherever they vary from QS-9000 requirements, customer-specific requirements prevail. Instead, QS-9000 is meant to be a *floor*, a basic set of generic requirements that can be applied to supplier/customer relationships among virtually all entities in the complex and dynamic automobile production system. As long as the product or service at issue is production material, production or service parts, or other specified services (Question 1), QS-9000 covers it. Neither the size of the supplier nor the location of the customer matters.

Who determines how QS-9000 is used? The customers, for the most part. All suppliers of the specified products or services are required to implement quality systems that comply with QS-9000. The exact timing varies with each customer, but, in some way, customers will verify that their suppliers are meeting the required standards. Several options for verification are possible:

■ The customer may audit the supplier directly, using the QS-9000 *Quality System Assessment* (QSA). This document (actually, a massive checklist) covers virtually all the "shall" statements of QS-9000. The customer may carry out an on-site audit. Or, certain suppliers may be told to audit themselves against the QSA and report on the results.

- The customer may require the supplier to obtain *third-party registration* to QS-9000. To meet this requirement, the supplier must contract with a registrar—an approved, accredited, independent company (Question 21). The registrar:
  - Audits the quality system against the QS-9000 standard.
  - Documents and reports any noncompliances.
  - Verifies resolution of the noncompliances.
  - Monitors the supplier twice yearly via "surveillance assessments," to make sure the requirements are continually being met.
- The Big 3 customers, as a rule, permit "registered" companies to be immune from duplicate or redundant "supplier quality assurance" audits, as long as they:
  - Meet customer-specific requirements.
  - Demonstrate continuous improvement.

QS-9000, then, can protect both customers and suppliers in the automotive supply chain from redundant, duplicative, and wasteful oversight activities. When the "fundamental" requirements are understood, agreed to, and (usually) confirmed by objective third-party audit, customers can be confident of the integrity and effectiveness of their suppliers' basic quality practices. The customer and supplier can then invest their energies and resources in agreeing to and working on the specific requirements that are unique to their relationship.

Does QS-9000 begin to sound like a mandatory, gun-to-your-head, my-way-or-the-highway program? For many suppliers, it is exactly that ("Get QS-9000 or get lost"). For many others, it is perceived that way. "They're trying to tell us how to run our businesses," the suppliers wail.

QS-9000 is not intended to strengthen the customers' control over how their suppliers run their businesses. The goal is to improve suppliers' quality and efficiency, resulting in cost savings throughout the supply chain. Admittedly, implementing QS-9000 does not guarantee these results. Like most things, the outcome depends on what is put into it. A compliant QS-9000 system can yield strictly cost and no benefit. This happens, usually, when the supplier tries to squeak by, do just enough to get registered, or "Get this thing done without changing how we work."

Suppliers who implement QS-9000 fully—in the spirit as well as the letter of the rules—can and do achieve the intended benefits. Perceiving this, some suppliers take the proactive step of implementing a documented quality management system (QS-9000 or ISO 9000) even if they are not being required to by customers.

4. What is QS-9000 "registration" or "certification" all about?

CAPSULE ANSWER

Registration, conferred by an independent, accredited third-party registrar, confirms that a supplier's quality system conforms to QS-9000 and communicates that conformity to others.

"Registration" is documented and objective evidence that a company's *quality system* meets the requirements of QS-9000. "Certification" usually applies to verification of the *quality of products*. In the context of QS-9000, the two terms often are interchanged and mean the same thing, but QS-9000 registration does not "bless" the quality of the product.

Why is registration so important? Two of the Big 3 are requiring registration for their direct suppliers of production materials, production or service parts, and heat treating, painting, plating, or other finishing services. Chrysler requires registration by July 31, 1997; for GM, the deadline is December 31, 1997. Ford is not requiring registration for the time being, but is encouraging its Q1 and other top-tier suppliers to demonstrate compliance with QS-9000.

Registration is important to the Big 3. It relieves them, in large part, of the need to verify that their direct suppliers are carrying out the fundamental quality practices spelled out in QS-9000. Verification is done, on the basis of thorough and impartial assessments of the quality system, by objective, third-party organizations called "registrars" (Question 21). These companies are:

- Wholly independent.
- Accredited by a recognized international accreditation body (Question 20).
- Approved by the Big 3.
- Selected and paid for by the would-be supplier.

Registration can cover:

- The sole location of a single-location company.
- All the locations of a multilocation company.
- Only certain units of a multilocation company (under specified conditions).
- Separate locations under separate certificates (a more costly approach).

The registration body audits a quality system against the requirements of QS-9000 and reports its findings in writing. These findings may (and usually do) include noncompliances (Question 96) that must be closed out prior to official registration. When the noncompliances have been corrected, the registration body:

- Lists the company's name in its book of registered companies (literally registers the company in its book).
- Issues a certificate of registration to the newly registered company. This certificate states all of the following:
  - Identity of the organization.
  - Location(s) covered by the registration.
  - A list of products/services supplied by the registered locations.
  - Revision date of the Standard.
  - Registration effective dates.
  - Name and location of registrar.

Most registrars limit registrations to three years. To renew the registration, another complete systems audit may be required. Some registrars have replaced the renewal audit with ongoing checks of the system via surveillance audits.

Under either scheme, to keep its registration, a company must undergo a surveillance assessment every six months. Surveillance assessments are scheduled events (there is no such thing as a "surprise" surveillance audit). Only part of the quality system is checked at each surveillance, and the registrar usually does not disclose the part that will be assessed until the day of the assessment. Over the three years of the registration, the entire quality system is checked in rotation via surveillance audits.

Companies can "fail" surveillance assessments or registration audits (Question 96). Normally, registrars allow a grace period, but the correction of noncompliances must be done within the agreed-on time in order to be granted registration or renewal. Registrars are required to report failures and "deregistrations" to the Big 3.

Registration is also available to Tier 2 suppliers (suppliers to Tier 1 companies). In fact, many Tier 1 suppliers are requiring key Tier 2 firms to register to QS-9000.

Each registrar publishes a list of the firms it has registered to QS-9000. A comprehensive list of QS-9000 registered firms is available via the American Society for Quality Control (ASQC), either directly or via the World Wide Web (WWW). (See Appendix C.) In addition, a list is compiled and published in subscription form by Irwin Professional Publishing (703-591-9008).

5. What firms are required to register to QS-9000?

CAPSULE ANSWER

Direct suppliers of production materials, parts, or key finishing services—and certain other suppliers—must register either to QS-9000 or to ISO 9000.

Three categories of firms, located anywhere in the world, are required, in some sense, to register to QS-9000:

1. Firms that supply the following products or services directly to Chrysler, Ford, and GM, or to any of the other original equipment manufacturers (OEMs) that recognize QS-9000:
   - Finishing services, including heat treating, painting, and plating.
   - Production materials.
   - Production or service parts.

   These firms are often referred to as Tier 1 suppliers.

2. Firms that are vendors to the Tier 1 suppliers. The Tier 1 suppliers may request and/or mandate that these Tier 2 suppliers or "subcontractors" implement QS-9000 systems and obtain registration. Chrysler, Ford, and GM do not, at this point, require Tier 2 companies to register to QS-9000. Consequently, not every Tier 1 company is requiring its suppliers to implement QS-9000; application varies from customer to customer. However, many Tier 1 companies, noting that Element 4.6 mandates "subcontractor development" to the QS-9000 standard (Question 43), are moving aggressively to get their Tier 2 suppliers into QS-9000. Their reasons vary, but the Tier 1 companies unquestionably have the authority to make that demand.

3. Facilities that make and/or distribute "regulated products"—products covered by directives issued by the European Union (EU) Council of Ministers. Sooner rather than later, these facilities must obtain ISO 9000 quality system registration in order to market those products in the EU. A company that makes, or is planning to make, one of these finished regulated products, or a component or element for one of them, should consider obtaining registration to:
   - QS-9000, if also a Tier 1 or Tier 2 supplier that is being required to register to QS-9000 by a customer.
   - ISO 9000, if not under a mandate to register to QS-9000.

Often, ISO 9000 registration is required of regulated products firms because they must also obtain product certification, which is a separate and huge issue beyond the scope of this book.

Directives for the following product categories have been issued by the EU and are being implemented now:

- Active implantable medical devices.
- Construction products.
- Electromagnetic compatibility.
- Machinery.
- Medical devices.
- Natural gas appliances.
- Nonautomatic weighing instruments.
- Personal protective equipment.
- Simple pressure vessels.
- Telecommunications terminal equipment.
- Toys.

Directives for the following product categories are proposed or planned for the near future:

- Amusement park and fairground equipment.
- Cable ways.
- Fasteners.
- Flammability of furniture.
- In vitro diagnostics.
- Measuring and testing instruments.
- Playground equipment (including sports equipment).
- Pressure equipment.
- Recreational craft.
- Used machinery.

Medical devices are examples of the product certification/quality system registration requirement. They have traditionally been controlled by various (and not always consistent) government regulations, because of the obvious risks involved in their use. The EU product directives cover some 150,000 medical products. All devices sold within the EU must meet the directives' requirements within five years of coming into force. They must also bear a "CE" mark as a symbol of conformity. The products are classified in one of four categories, according to the degree of risk inherent in using them. Each classification is *as relaxed as possible* to ease bureaucratic burdens on business, and *as strict as necessary* to ensure that patients' health is adequately protected. Once a medical product and its associated quality system are assessed, certified, and registered, the certification and registration are recognized throughout the EU.

To do business in the EU, product manufacturers need to conform to just one standard.

6. What are the advantages or benefits of implementing a QS-9000 quality system?

> **CAPSULE ANSWER**
>
> QS-9000 creates consistency, provides continuous improvement, reduces dependence on individuals, and provides the structure a company needs to react to change.

Most companies get involved in QS-9000 because "our customers are making us register." In their view, that is the sole benefit: Keeping their customers happy.

A company that gets into QS-9000 for that purpose alone, without pursuing it for the other benefits it provides, is truly losing out. It is setting itself up to be in the worst possible position: QS-9000 is a cost, not a benefit.

What can a well-implemented QS-9000 system do?

- Create consistency throughout a company. Consistent working methods and quality controls are established and enforced. In large, multi-site organizations whose facilities are major suppliers to each other, this consistency can be especially important.
- Strengthen relationships between both internal and external suppliers and customers. As a documented quality system, especially with its insistence on cross-functional decision making, QS-9000 is common ground for addressing quality issues of mutual importance.
- Provide customers with confidence in the capability of a company to live up to its quality commitments. This benefit is much stronger when the quality system is registered.
- Improve management decision making. A quality system is an information system. Internal audits, management reviews, analysis of company-level data, and effective document and data control—the pillars of QS-9000's strength—provide management with the intelligence needed to make the right moves.
- Facilitate continuous improvement. The documented continuous improvement system, plus the corrective and preventive action activities required, instill an attitude of prevention rather than detection.

- Institutionalize training in methods and procedures that are essential to quality.
- Reduce dependence on individuals. People are vital to quality, but people come and go. The levels of procedural development, documentation, record keeping, and training required by a QS-9000 quality system ensure that techniques and skills will carry on even when performed by different individuals.
- Convey a sense of accomplishment. QS-9000 works—not overnight, not without pain, and not as a panacea, but it works.
- Add value. QS-9000 is too new for there to be any definitive studies on its value. For ISO 9000, though, the evidence is clear. Facilities with advanced cost-tracking controls almost always find that their documented quality system adds value. A major home appliance manufacturer saw its failure rate (claims per year divided by sales per year) drop by 70 percent in three years. Its warranty cost per unit declined by 76 percent during the same period.

Dupont, a pioneer in quality improvement and in QS-9000 implementation, had several measures for the improvements realized under QS-9000:

- On-time delivery increased from 70 percent to 90 percent.
- Cycle time improved from 15 days to ½ day.
- First-pass yield improved from 72 percent to 92 percent on a product line.
- One site reduced its more than 3,000 test procedures to 2,000.

Lloyd's Register Quality Assurance, the British quality assurance registrar, published a survey of some 400 of its ISO registrants in the United Kingdom. The population was a proportional sample of market sectors and company sizes. Here are some of the findings:

- 67% felt that the QS-9000 approach was essential for creating and maintaining viable quality management systems.
- 69% reported that QS-9000 improved productivity and efficiency.
- 73% felt that QS-9000 quality systems ensured their products' consistency and enabled them to deliver better service to their customers.
- 86% stated that their QS-9000 systems improved management control.
- 89% agreed that the internal benefits of QS-9000 "met or exceeded expectations."

■ Most respondents had originally sought ISO for external benefits, but discovered that the internal benefits were more beneficial.

Perhaps the most succinct assessment of improved operations under an ISO 9000 quality system has come from Phil McNamara, regional quality manager for Bass Brewers (a U.K. company), which registered 39 sites employing 4,500 persons: "ISO 9000 has given us a much more structured approach to change. . . . It has taken a lot of effort and commitment, but the payoff is the elimination of a lot of duplicate effort and waste."

7. What are the advantages or benefits of registration to QS-9000?

Implementing a QS-9000 quality system brings significant benefits to a company (Question 6), whether it is registered or not. Registration leverages those benefits.

For one thing, registration itself becomes an initial goal. As implementa-

> ### CAPSULE ANSWER
>
> QS-9000 registration improves customer confidence, provides access to markets, improves competitive standing, and reduces the costs of supplier quality assurance programs.

tion moves along, more and more employees get drawn into the cooperative effort. By the time of the registration audit, virtually everyone in the company is aware of being a participant. And when the effort of many months is paid back by "passing" the registration audit, the entire company experiences a real morale boost. But registration is not the checkered flag. It is the green flag that triggers greater efforts.

QS-9000 registration does not simply allow a company to say, "Great, now we can keep supplying Ford." It brings these other key benefits, too:

1. Gives customers confidence that a firm can meet its quality commitments. Customers don't have to do an audit themselves or take the supplier's word for it. The judgment of a qualified, objective, third-party registrar is restated every six months.
2. Provides access to markets. Most companies registering to QS-9000 today are doing so because key customers (OEMs or Tier 1 firms) are pressuring them to register. Yet, the huge QS-9000 market is just one market. Other markets that put great store in QS-9000/ISO 9000 registration should not be ignored. Here are some of the reasons:

- A company that makes and/or markets products covered by European Union (EU) product directives (Question 5), or plans to do so in the future, may be compelled to register its quality system in order to operate in the EU countries.
- Other marketplaces may become less and less friendly to unregistered firms as the number of registrations increases.
- In a recent survey, more than 80 percent of respondents said ISO 9000 registration would favorably influence their choice of suppliers.
- Many companies are finding that ISO 9000 or QS-9000 registration is often an item on supplier surveys.

3. Reduces cost of customers' supplier programs. To the extent that customers accept QS-9000 registration in lieu of supplier quality assurance audits, their own costs go down. (This is clearly one of the motivations of the Big 3.)

4. Reduces suppliers' operating costs. British Standards Institution (BSI), possibly the world's largest and most respected quality assurance registration body, estimates that registered firms reduce their operating costs by 10 percent, on average. However, a system's contribution depends largely on the company's starting point. If operation is already at a peak of efficiency, QS-9000 registration is not going to pay back another 10 percent.

5. Provides competitive advantage. QS-9000 registration is a powerful marketing tool. Registered firms can proudly display their certificate and logo, and their names appear in registries of approved firms. Quality is a strong differentiator in parity markets. QS-9000 registration is objective, confirmed evidence of an active, thriving quality system.

6. Reduces supplier quality assurance (SQA) audits. Some companies are subject to as many as 30 to 40 supplier quality audits a *month!* As QS-9000 gains visibility and credibility, registration is increasingly easing acceptance to approved supplier lists. In some cases, it eliminates supplier audits entirely.

The creators of QS-9000 intended for the quality system standard to return solid benefits to all concerned:

> These companies [the OEMs] are committed to working with suppliers to ensure customer satisfaction beginning with conformance to quality requirements, and continuing with reduction of variation and waste to benefit the final customer, the supply base, and themselves.

The creators of QS-9000's parent, ISO 9000, cited a more specific benefit of registration:

> An important corollary benefit for any organization is reduction of the costs of multiple assessments by multiple trading partners. In practice, purchaser organizations often audit portions of the quality systems of their suppliers, but because of supplier quality system certification, the purchaser does not have to duplicate the, say, 80 percent that has already been audited by the third-party auditor.

8. What is the cost of registering to QS-9000?

For those who turned to this question first, the short definitive answer is: It depends.

Two kinds of costs are involved here:

> **CAPSULE ANSWER**
>
> The cost of QS-9000 registration depends on many factors: size of a company and number of locations; marketplace forces; and price competition among registrars.

1. The cost of implementing the system.
2. The cost to engage the registration body not only for the registration audit itself, but also for associated activities: preassessment, surveillance assessments, and so on (Question 4).

IMPLEMENTATION COSTS

Time, energy, and physical resources are needed to set up a QS-9000 system and prepare for registration audit. They all translate into money being spent. However, once a QS-9000 system is implemented and has reached steady state (Question 100), operation of the system should not increase overhead.

How much does it cost to implement the system? So many factors are relevant, it is impossible to state a meaningful dollar figure here. Head count is a big factor. The more people on board, the more training must be done. Beyond personnel training costs, there are other significant factors.

Implementation will tend to cost more if a company:

- Has more than one location.
- Is responsible for design.
- Has no active quality system now.
- Is undergoing any kind of significant corporate change, such as:
  - Downsizing.
  - Chapter 11 reorganization.
  - Merger/acquisition.
  - Implementing a significant new process.
  - Implementing a new EDP system.
  - Relocating/reconfiguring.

Implementation will tend to cost less if a company:

- Is already registered to ISO 9000 (Question 2).
- Is a Pentastar, Q1, or Targets for Excellence company. (But not necessarily; see below.)
- Dedicates a seasoned, responsible manager to championing the effort. (This person would probably become the Management Representative.) Once registration is achieved, he or she would go back to prior job duties, inasmuch as the QS-9000 responsibility would, at best, be 25 percent of his or her job.
- Hires good consultants.

The last item usually produces an echo: "Hiring a consultant *saves* money?" Odds are, yes—if a *good* consultant is hired (Question 80).

Regarding Pentastar, Q1, and Targets for Excellence, these prestigious certificates tend to suggest that many of the practices required by QS-9000 are already in place. Still, implementation will not be easy, nor will registration be a given, for two reasons:

1. QS-9000 has several significant requirements, inherited from ISO 9000, that are not mandated by Pentastar, Q1, or Targets for Excellence. Implementing them will take time and money.
2. Sad to say, some companies have received Pentastar, Q1, and Targets for Excellence designations by virtue of audits that were not overly rigorous. (I have personally seen more than one Tier 1 supplier, displaying all the proud banners and certificates the Big 3 could bestow, whose gage calibration systems were, upon close examination, what

are called in the trade "sticker and snicker" systems. They do not fly with QS-9000.)

Estimating the costs of the following activities, which are directly related to the implementation, is fairly easy:

- Overview training—30 minutes of time for every employee in the organization.
- Orientation training—perhaps 90 minutes of time for every employee in the organization. (This estimate is a hard one to call.)
- Documentation writing training—two workdays of time for perhaps 12 to 15 key operations people from a cross-section of the organization.
- Internal quality audit training—two workdays of time for a group of employees that represents about 10 percent of the total head count.
- Internal quality audit costs (this estimate is very general)—an average of four hours per audit for two auditors (total of eight hours per audit), times the number of standard operating procedures (SOPs) in the system (at least twenty but could be as many as twenty-six). At least one complete cycle of internal quality audits must be completed before the registration audit.
- Equipment, supplies, and so on, including a good computer with word processing software and the services of someone who knows how to use it.

Other costs are almost impossible to estimate:

- Management representative's (MR) time. The typical MR does not work on the project full time. He or she usually has other responsibilities. Many MRs do much of their QS-9000 work on an overtime basis, especially during implementation.
- Document review time. Managers and others need to review SOPs and other documents, make notes, suggest changes, and so on.
- Corrective/preventive action activities (another estimate that is hard to call). If a company already has such activities in place, then QS-9000 will not add a lot of time. If the company's activities in this regard are informal and hit-or-miss, the activity will take the time of key people, especially while they are learning how the system works.
- Document control activities. This burden usually falls on the MR and/or his or her staff. When the system is set up and is running smoothly, minimal time is needed—as long as the document system is

kept as lean as possible! (Anyone who lets a document system bloat out to dozens and dozens of SOPs will find no sympathy here.)

REGISTRATION COSTS

These costs can vary also, but at least they are a bit easier to contain. Quotes from a number of reputable registrars (Question 91) have to be gathered, analyzed, and compared. The caveat emptor principle applies.

Registrars usually price their services on a sliding scale governed by three factors:

1. Design responsibility.
2. Number of locations (if a multisite registration).
3. Size of facility in terms of number of employees. This translates into a required number of audit days as stated in QS-9000 and published in a schedule (Question 15).

In actual practice, registration costs can vary dramatically. Registrars have different "daily rates." Some have application fees, or annual administration fees, and some do not. Turning competing registration quotes into "apples and apples" can be an exercise in and of itself.

Here are some documented examples of the range of the total cost of a three-year registration. In both cases, five accredited and approved QS-9000 registrars submitted bids based on the same information:

■ A Tier 1 manufacturer with two manufacturing sites, a satellite warehouse, full design responsibility, and about 400 employees: $41,000 to $60,000.
■ A Tier 2 manufacturer with one site, no design responsibility, and about 100 employees: $11,500 to $24,000.

These are real numbers quoted during 1996. They should be used only as very rough guides. Also, prices are declining because QS-9000 registration is, in large part, a market-driven process. Only certain registrars are authorized to issue QS-9000 certificates, but within that universe of some 20+ registrars, competition on price is intense. Applicants are free to negotiate price, and negotiation is strongly recommended. However, price alone should never be the determining factor in selection of a registrar (Question 91).

9. How long does it take to register to QS-9000?

Like the cost of registration (Question 8), the time needed to become registered varies. The process will take months, and a business cannot simply shut down while getting registered.

The entire process can be broken down into the following general phases:

> **CAPSULE ANSWER**
>
> Registration time depends on the state of the quality system at inception, as well as other factors. Generally, the registration process can take between eight and twenty-four months to complete.

- Implementing the QS-9000 system (see below).
- Operating the QS-9000 system for the minimum time (three, or preferably six, months before a registration audit).
- Selecting a registrar. This can be ongoing during the first two phases, to save time.
- Awaiting a registration audit, after application. The length of this interval depends on the registrar's backlog. In late 1996 and early 1997, registrars are expected to be swamped, and lead times may become extended.

The time it takes to implement the QS-9000 system depends in large part on a company's starting point. A company that has any of the following should experience a relatively short implementation time:

- A registered ISO 9000 quality system.
- Pentastar, Q1, or Targets for Excellence credentials. (But not necessarily; see Question 10.)
- A documented quality system of any kind that is active and meaningful but not necessarily compliant with any particular standard.
- Resources temporarily dedicated solely to implementing the system.
- The guidance of a *good* consultant (Question 80).

For a company starting from square one, implementation will take a long time. Other factors that can extend the implementation time include:

- Number of locations.
- Head count.
- Design responsibility.

- Corporate turmoil.
- Lack of an ongoing and steady top-management commitment, exhibited in:
  - Lack of sufficient resources.
  - Priority given to other issues.
  - Vacillation.
  - Failure to pay attention.
  - Failure to learn and understand.
  - Failure to lead

In my experience, the bottom line is:

- On average, the shortest interval for the entire process—from launch through registration audit—seems to be 6 to 9 months.
- With a large, multisite, design-responsible company, I've seen the process take 18 to 24 months, even with significant resources and full management commitment.

On average, a company will be looking at about a year to get the registration done.

10. If my company has Pentastar, Q1, or Targets for Excellence now, how hard can it be to get registered to QS-9000?

With any of those certificates, registration is certainly easier than implementing a QS-9000 system completely from scratch. But the certificates will not necessarily guarantee an easy ride to registration.

> **CAPSULE ANSWER**
>
> If you have earned a Pentastar, Q1, or Targets for Excellence certificate, congratulations. But having these credentials does not guarantee an easy ride to QS-9000 registration.

There is a story, perhaps apocryphal, about a Tier 1 company that had Pentastar, Q1, and every other certificate available. When QS-9000 was officially released in the summer of 1994, this company decided to become the very first to register to QS-9000. It rushed through a cursory implementation, brought in the registration assessors—and got blown out of the water.

This story may be a fable but, like most fables, it has a moral. "Quality flags" should not lull a company into a feeling of complacency about the ease with which QS-9000 implementation and registration can be accomplished. Here's why:

■ *QS-9000 is based on ISO 9000, which has its own specific document requirements.* This is not to say that applicants must trash their existing documents and start over. Far from it (Question 82). But odds are that any system will have to be reexamined and largely restructured.

■ *QS-9000 has requirements not found in the automotive quality programs.* The need for management review, a management representative, contract review, document control, and other elements may be new to many QS-9000 candidate firms. It takes time and effort to implement these systems effectively.

■ *QS-9000 requires a generally higher level of awareness throughout the organization.* This is not a "quality program" left to the "quality department" to handle. It requires the involvement and awareness of virtually everyone in the company.

■ *QS-9000 is, in general, more rigorously audited.* Audits are a touchy subject. There are certainly exceptions, but, in general, audits carried out by the independent, third-party QS-9000 registrars (Question 21) are more thorough, comprehensive, and rigorous than the typical "supplier quality assurance" second-party audits carried out in the past. Not all second-party audits were "wink wink, nod nod" affairs, and not all QS-9000 audits have the level of rigor and thoroughness expected by the Big 3. But the fact that the registration audits are done by third parties tends to increase the level of compliance expected.

Being a Pentastar, Q1, or Targets for Excellence company is certainly something to be proud of. With the related systems in place and well implemented, the path to QS-9000 registration will certainly be easier.

But the other systems must be put aside to allow full *focus on the QS-9000 requirements.* These steps are recommended:

■ Do a self-audit. (The *Quality Systems Assessment* checklist is excellent for this.)
■ Make an honest and realistic list of any "compliance gaps."
■ Develop the systems needed to eliminate those gaps.
■ Implement the QS-9000. (See Part Five for more details.)

In large part, it is not necessary to "reinvent the wheel." The Production Part Approval Process (PPAP) systems currently in use will work just fine under QS-9000. So will the Failure Mode Effects Analyses (FMEAs), control plans, and gage repeatability and reproducibility (R & R)—assuming that the present systems meet customers' requirements and no corners are being cut. The more defined and disciplined the quality practices now in place, the easier the path toward reaching QS-9000 registration. But all the Pentastar, Q1, and Targets for Excellence certificates ever printed will not, in and of themselves, guarantee a painless and changeless trip.

11. We're leaders in our market. Our customers love us. Our quality is unquestioned. What does QS-9000 offer us that we don't have already?

> **CAPSULE ANSWER**
>
> QS-9000 can help your company maintain access to key customers, improve performance, and achieve a new level of international credibility.

That's not a question. It's a speech that gets made fairly regularly, so let's address it.

What does QS-9000 offer? To a company that is required to register (Question 5), it offers continued access to existing customers. That in itself is a pretty strong benefit.

Customers may in fact love a company's work and never question its quality. But any customers—the Big 3 and others—will depend most heavily on their QS-9000 Tier 1 suppliers. The Big 3's vested interest in improving their suppliers' performance is based on some facts of industry life:

- Being a "great" company does not mean it is as great as it could be. QS-9000 mandates a "continuous improvement" system. By implementing that system and working within it conscientiously, a company cannot help but improve.
- Being "great" today does not guarantee being great tomorrow. Is the industry changing? Has the company changed? A well-implemented QS-9000 system helps with adapting to change. It brings independence to individuals and consistency to practice—two features that tend to resist any decline in performance.
- Any suppliers not interested in meeting the QS-9000 requirements are free to walk away.

What else does QS-9000 bring? When well implemented, the system can improve a company's performance, which is, after all, the whole point. When no improvement occurs, the fault tends not to lie in the QS-9000 system, flawed though it is known to be. Instead, when a QS-9000 system does not provide substantial benefits and improvement in performance (Question 6), it is usually because management has consciously chosen to cut corners, blow smoke, stay uninvolved, and starve the system of all but the most essential resources. "We'll do this stupid thing, but we're sure not going to change the way we operate."

QS-9000 registration brings one more thing that a company may not have today: international credibility. QS-9000 is based on ISO 9000 (Question 2), the internationally recognized quality system standard. Registration to QS-9000 includes registration to ISO 9001 or ISO 9002—in a way, two registrations are gained for the price of one! This credential may gain the attention of automakers overseas. The Big 3 are indeed big, but they are not the whole ballgame.

12. Should my company or organization seek QS-9000 registration now?

Sometimes, this is a very easy decision. For a provider of a certain category of products or services directly to the Big 3 (see the chart on pages 27–28), the choice is simple: Get registered, or walk away from the business.

The situation gets a little stickier for firms that supply something other than those categories of products or services, either directly to the Big 3 or to one of the Tier 1 suppliers. For those firms, ISO 9000 (rather than QS-9000) might make more sense.

The chart on pages 27–28 is a quick decision-making tool.

13. How can we be sure that our QS-9000 system won't just fade away?

QS-9000 is not a "program of the month"—here today, with much sound and fury, and, after a slow, embarrassing fadeaway, gone tomorrow.

CAPSULE ANSWER

A QS-9000 system will not fade away as long as management sees value in remaining registered.

Registration Mandate	Circumstances	Comments
Registration to QS-9000 mandatory IF:	▪ A company supplies finishing services (heat treating, painting, plating, or other services), production materials, or production or service parts directly to Chrysler, Ford, GM, or other OEM sponsors of QS-9000, and wants to keep their business ▪ One of the company's customers supplies the products or services listed above directly to OEM sponsors of QS-9000 and wants to keep their business. ▪ One or more direct competitors is presently registered. ▪ One or more important customers is presently registered.	Under these circumstances, registration is generally mandatory in order to (a) maintain and expand access to vital markets, and (b) protect a competitive position.
Registration to ISO 9000 mandatory IF:	▪ A firm markets a "regulated product" (Question 5) internationally.	Under there circumstances, registration is generally mandatory for the reasons given above.
Registration to ISO 9000 highly advisable IF:	▪ A company supplies something *other than* finishing services, production materials, or production or service parts directly to the Big 3. ▪ A company supplies something *other than* finishing services, production materials, or production or service parts to a direct supplier to the Big 3.	QS-9000 is most likely not appropriate, but customers may eventually insist on implementing a documented quality system and registering it to ISO 9000. The customer may be pressured to register and may in turn pressure the next tier of suppliers.

(Continued)

Registration Mandate	Circumstances	Comments
Registration to ISO 9000 optional IF:	■ A corporate parent and/or corporate affiliates are registered to ISO 9000	Such companies almost inevitably find that the decision to register has been made for them.
	■ A firm has a well-designed and efficient quality system.	ISO 9000 registration might make sense simply to keep "ahead of the curve" and/or to increase competitiveness. If the current quality assurance system is really good, conformity and registration to ISO 9000 should not present tremendous problems.
	■ One or more competitors is rumored to be seeking registration.	Probably, it would be wise to make the move soon, in order to stay competitive.
	■ One or more RFQ's has specified QS-9000 registration as a quality assurance option. ■ A firm markets a non-regulated product internationally. ■ A major corporate goal is to become more competitive.	
	■ A firm lacks any sort of quality program or system.	ISO 9000 is an ideal place to start.
Delay registration IF:	■ None of the above is true—and a firm struggling to survive.	A struggling organization without a quality system and without the competitive necessity to implement one should focus on fixing its business problems. Certain parts of an ISO 9000 quality system can (and should) be implemented on a gradual basis.

The QS-9000 system won't fade away as long as top management remains committed to it. And top management will remain committed to it as long as it is returning one of these benefits:

■ Current business stays.
■ New business comes.
■ Incremental cost savings are realized.

Because most companies get into QS-9000 as a result of customer pressure, the first benefit listed above is the most operative one. The second benefit is speculative and, at this early stage of the QS-9000 era, unproven. The third benefit, surprisingly, is genuine: ISO 9000 registrants, with virtually no exception, realize proven cost savings within about two years.

Top management will stay committed to the system in order to maintain existing business and in the hope of obtaining new business. The company must therefore remain registered, which means it must undergo and pass surveillance assessments every six months. Of the three "reinforcement mechanisms" of QS-9000—the attributes that keep the system from fading away as another "program of the month"—the surveillance assessments are probably the most potent.

The second reinforcement mechanism, the management review process (Question 31), requires senior management to do scheduled reviews of the QS-9000 system from top to bottom: its implementation, suitability, effectiveness, and results. Records must be kept as proof that the reviews are done. Besides forcing management to pay atten-

THE REINFORCEMENT MECHANISMS
1. Surveillance assessments
2. Management reviews
3. Internal quality audits

tion to the system, the reviews create an educational process for management. Over time, they show how useful the QS-9000 system can be as a management and communications tool.

The third reinforcement mechanism is the required internal quality audit process (Question 32). Trained, independent employees audit the entire quality system on a scheduled basis and record the results. Corrective actions against deficiencies found during these audits must be carried out and verified. Internal quality auditing is an outstanding implementation tool (Question 88). It also keeps the entire organization tuned in to the system, improving it as an ongoing activity.

To remain registered, a company must pass surveillance assessments and work the entire system, including management reviews and internal quality audits. Once the system matures and becomes transparent, QS-9000 ceases to be a "program." It becomes "the way we do things around here."

PART TWO

Origins, Elements, and Applications of QS-9000

14. How did QS-9000 develop?

QS-9000 is an effort to "harmonize," or make consistent, the quality system requirements imposed by the sponsoring customers (Big 3 automakers and others) on their main suppliers.

For many years, each of the Big 3 had required its main suppliers to comply with its own set of quality system requirements:

CAPSULE ANSWER

In structure and (to a large extent) content, QS-9000 is based on the international ISO 9001 standard which evolved from a long series of quality system standards.

- Chrysler: Pentastar.
- Ford: Q101/Q1.
- General Motors: Targets for Excellence.

Many aspects of these requirements were similar. Others were contradictory. Usually, customers enforced compliance with its own schedule of audits. So, for companies that supplied more than one Big 3 customer, life could be difficult. They had to meet two or more sets of conflicting and/or redundant requirements and deal with more than one set of auditors.

At the same time, things were not roses for the Big 3, either. In their intensely competitive market, cost control and cost reduction became watchwords. Their systems for dealing with suppliers came under scrutiny. It became obvious to all concerned that harmonizing the quality requirements made sense. If they could do this, suppliers would only have to deal with one set of rules and one set of audits. And the customers would greatly reduce their costs of monitoring and enforcement.

So the Automotive Industry Action Group (AIAG) (Question 18) got the assignment of creating a harmonized quality system standard. They had to rationalize the somewhat different Big 3 systems into one streamlined and (hopefully) workable system. The resulting system had to have international standing and credibility.

Enter ISO 9000. In a move that was shrewd indeed—and could very well prove to be the stroke that ensures that QS-9000 will ultimately work—the AIAG (along with the Big 3):

- Adopted the structure of the main ISO 9000 standard (ISO 9001).
- Included the ISO 9001 requirements word for word into QS-9000 (adding harmonized and more prescriptive automotive-based requirements).
- Incorporated the process of third-party registration (Question 4), using existing registration bodies that have international standing and credibility (Question 21).

QS-9000 is based in large part on the requirements of ISO 9001. This standard, published by the International Organization for Standardization (ISO) (Question 19) is much more generic. It sets a benchmark for the quality systems for organizations of virtually any type, producing any kind of product or service, anywhere in the world.

By basing QS-9000 so directly on ISO 9000, the U.S. automotive industry tacitly acknowledged the existence and credibility of the international standard, rudimentary though it is. It is also fair to say that some aspects of QS-9000 are generic enough to merit consideration as additions to ISO 9001. These aspects, which are absent from ISO 9001, include:

- A documented continuous improvement process.
- Customer satisfaction determination process.
- Evaluation of training effectiveness.
- Strengthened statistical analysis.

Just as QS-9000 did not emerge from a vacuum, nor did ISO 9001. First published in 1987 (and modestly updated in 1994), ISO 9001 was based on a series of quality standards published in the United Kingdom in

1979. This series, known as BS-5750 (BS stands for British Standard), was sponsored by the U.K.'s Department of Trade and Industry. The program, which included third-party audits and registration of firms that met the Standard, was widely publicized to British businesses and consumers in an effort to raise the level of awareness, and fulfillment, of quality. Because BS-5750 was so aggressively pushed by the government, it was adopted by a wide range of organizations well beyond the traditional manufacturing sector.

BS-5750 was not invented overnight. Its evolution can be traced backward, in a sort of family tree, to a series of military quality standards:

- Def Stan 05-20 Series (UK Ministry of Defence).
- AQAP-1 (NATO, 1968).
- MIL-Q-9858 (United States, 1958).

An ancestor of today's QS-9000 is the venerable (and, until recently, still widely used) American defense quality standard MIL-Q-9858. What goes around does, indeed, come around.

15. What documents directly relate to the QS-9000 requirements?

The main document is *Quality System Requirements QS-9000*. This 108-page book, published by the Automotive Industry Action Group (AIAG) (Question 18), spells out the QS-9000 requirements. It also documents many of the other "rules of the road" for the various parties involved in QS-9000: suppliers, auditors, registrars, and accreditation bodies.

> **CAPSULE ANSWER**
>
> The main documents are *Quality System Requirements QS-9000* and *Quality System Assessment* (QSA).

These requirements are supplemented from time to time by a document called *IASG Sanctioned QS-9000 Interpretations*. Prepared and periodically updated by the International Automotive Sector Group (IASG), this document provides clarifications and interpretations of QS-9000 requirements and other matters related to the QS-9000 system. It is available from the American Society for Quality Control (ASQC).

The other principal document is *Quality System Assessment* (QSA), chiefly a detailed checklist of QS-9000 requirements. Used as an auditing tool by internal, second-party, and third-party (registration) auditors, it includes:

- Instructions for the assessment process.
- Rules for the evaluation process.
- Sample forms.

Before beginning a QS-9000 implementation, a company will need the current versions of each of these documents. Getting them is the easy part. Reading them and developing an understanding of them requires some quiet and uninterrupted study time. The *ISO 9000 Standards Compendium*, available from the American National Standards Institute (ANSI), also provides a wealth of guidance and information on the ISO 9000-based requirements found in QS-9000.

QUALITY SYSTEM REQUIREMENTS QS-9000

The book's introduction defines the goal, purpose, and approach of QS-9000, and illustrates the "typical" QS-9000 document structure.

Section I itemizes the ISO 9000-based requirements. It is divided into 20 subsections, each representing one of the 20 elements of ISO 9001.

Each subsection reproduces (in italics) the exact language of the ISO 9000 requirement. Most of the sections also include (in roman type) additional, automotive-related requirements (Question 2). These distinctions in wording apply:

- Statements that use "shall" are requirements that must be adhered to.
- Statements that use "should" represent a preferred approach. Alternatives are acceptable as long as assessors feel that the intent of the statement has been met. (This elasticity can lead to some fairly interesting "judgment calls" on the part of suppliers and assessors.)
- "Notes" are for guidance only; they are not auditable.

Section II spells out "Sector-specific Requirements"—requirements not found in ISO 9001. The topics within this section are:

- Production Part Approval Process (PPAP) (Question 39).
- Continuous improvement (Question 27).

■ Manufacturing capabilities:
  – Facilities, equipment, and process planning and effectiveness.
  – Mistake proofing.
  – Tool design and fabrication.
  – Tooling management.

Section III details "Customer-specific Requirements," with a separate part for each major customer: Chrysler, Ford, GM, and the truck manufacturers. The requirements spelled out by each customer must be met by direct suppliers.

The balance of the book contains the following appendixes:

■ *Appendix A: The Quality System Assessment Process.* Defines the various types of verification available: second-party or third-party (quality system registrar) audit. Includes a description of the customer's decision-making process as to the type of verification to require.

■ *Appendix B: Code of Practice for Quality System Registrars.* The purpose as stated in the title; explained in detail in Question 93.

■ *Appendix C: Special Characteristics and Symbols.* Defines the various "special characteristics" (and related symbols) used by each of the Big 3 (Question 49).

■ *Appendix D: Local Equivalents for ISO 9001 and 9002 Specifications.* Lists the names for ISO 9001 and ISO 9002 in various countries. Includes the contact information for local standards bodies.

■ *Appendix E: Acronyms and Their Meanings.*

■ *Appendix F: Change Summary.* Identifies the changes made between editions of the Standard.

■ *Appendix G: November 21, 1994 QS-9000 Accreditation Body Implementation Requirements.* Contains special rules that apply to accreditation bodies (Question 20), the groups that "approve" or "sanction" quality system registrars. QS-9000 certificates can only be issued by quality system registrars accredited by Big 3-approved accreditation bodies. Big 3 approval is contingent on accreditation bodies' agreement to follow the Requirements in this section.

■ *Appendix H: Survey Audit Days Table.* States the number of audit days required for initial and surveillance audits, depending on the head count at the site. Subsequently modified by *IASG Sanctioned QS-9000 Interpretations.*

■ *Glossary.*

16. What documents provide additional guidance and recommendations for firms wishing to register to QS-9000?

Quality System Requirements QS-9000 spells out the requirements as well as information on registration. A number of other documents in both the QS-9000 family and the ISO 9000 family are worth studying while planning and carrying out the implementation.

> **CAPSULE ANSWER**
>
> Several "guidance" documents in the QS-9000 family have, by reference, the force of requirements in many cases. Some documents from the ISO 9000 family are also very helpful in interpreting and implementing the requirements.

Unless otherwise noted, the following QS-9000 documents are available from the Automotive Industry Action Group (AIAG; Question 18 and Appendix C).

1. *IASG Sanctioned QS-9000 Interpretations.* This document, updated several times a year, explains the QS-9000 requirements and clarifies application, registration, auditing, and other issues. Its provisions have the force of requirements, and it is helpful in applying the Standard in specific situations. Available from the American Society for Quality Control (see Appendix C).
2. *Quality System Assessment* (QSA). This is a checklist of questions related to the QS-9000 requirements. Registration auditors are required to audit all items on this checklist. The book includes definitions and explanations of audit scoring methods and is used for supplier self-assessments.
3. *Advanced Product Quality Planning and Control Plan Reference Manual.* This book is referenced under Element 4.2 (Quality System) of the requirements, which states that the book must be "utilized" in an advanced quality planning process. So it is a must.

   The book spells out a standardized product quality planning system applicable to automotive suppliers, whether design-responsible or not. Covered in detail are the following:
   - Program planning.
   - Design issues (design Failure Mode and Effects Analyses (FMEA), design reviews, special product and process characteristics, and so on).

- Process design and development, including flow charting, process FMEA, control plans, measurement systems analysis (gage R & R; see below), and preliminary process capability studies.
- Product and process validation, including process capability studies, production part approvals, and production control planning.
- Feedback, assessment, and corrective action.
- Control plan methodology

For firms that have already defined an advanced product quality planning process, this detailed manual offers many ideas for improvement. For those that have no system, the book includes virtually all the tools needed to implement one.

4. *Potential Failure Mode and Effects Analysis (FMEA) Reference Manual.* This book gives detailed instructions, guidance, and examples on developing and using design FMEAs and process FMEA. Examples and sample forms are included.

5. *Production Part Approval Process* (PPAP). Section 2.1 of the QS-9000 requirements describes production part approval (Question 39). The requirements apply whether a company is design-responsible or not. The PPAP manual details when PPAP is required, the various submission levels, and descriptions of the different elements of a submission. It also provides sample forms and additional customer-specific instructions (very important).

6. *Measurement Systems Analysis.* Element 4.11 of the QS-9000 requirements mandates that the measurement systems used to check quality characteristics must be statistically verified. These studies, commonly called gage R & R (repeatability and reproducibility) studies, are described in detail in this manual.

Some guidance documents from the ISO 9000 family are very helpful in interpreting and implementing the ISO 9000 portions of the QS-9000 standard. The QS-9000 books do not, in general, interpret or refine the ISO 9000 portions; this is left to the ISO 9000 manuals, of which the most important are listed here.

1. ISO 8402: *Quality Management and Quality Assurance—Vocabulary.* This glossary is a very carefully written and organized compendium of terms used in quality systems generally, and in ISO 9000 specifically. Many important terms are defined in the opening sections of the various ISO 9000 documents, but ISO 8402 is the single and ultimate

authority on the precise meanings of terms found in the Standard. The glossary is divided into the following sections:

- General Terms.
- Terms Related to Quality.
- Terms Related to Quality System.
- Terms Related to Tools and Techniques.

2. *Quality Management and Quality Assurance Standards—Part 2: Generic Guidelines for the Application of ISO 9001, ISO 9002, and ISO 9003.* The structure of this document is consistent with the paragraph sequence of QS-9000, Elements 1 through 20. Each section explains in some detail what is intended by the requirements. Though no additional requirements are found in ISO 9000-2, it contains many recommendations that the user is expected to consider when developing and implementing a quality system.

3. ISO 10011: *Guidelines for Auditing Quality Systems.* This three-part document discusses auditing, qualification criteria for quality system auditors, and management of audit programs.

17. What quality tools and techniques are mandated by QS-9000?

QS-9000 requires the use of a number of quality tools and techniques. These include techniques used in the design of processes and/or products, people-oriented methods, and statistical methods.

> **CAPSULE ANSWER**
>
> QS-9000 requires the use of mistake-proofing methods, disciplined problem-solving methods, and cross-functional teams for decision making.

Specifically, QS-9000 requires the use of "cross-functional" teams, and related methods, for decision making and in preparatory stages of production of new or changed products.

Disciplined problem-solving methods are required to attack nonconformities. A company must use format(s) that are approved by its customers.

In the design of products (applies to design-responsible companies only) and processes (applies to everyone), a method called "mistake proofing" must be used to prevent nonconformities. Mistake proofing, alias "poke yoke," is a process of implementing product (or process) design features that literally prevent mistakes from being made either during manufacture or while the product is in use. A number of "continuous improvement" methodologies are also required, as appropriate. (More on this below, and in Question 27.)

ISO 9000 also provides a guidance document on quality improvement methods. The document outlines details on the use of numerous traditional quality improvement tools and techniques.

TECHNICAL REQUIREMENTS

1. A company must use a cross-functional, multidisciplinary approach for decision making and when preparing to produce new or changed products. Functions to be included are:
   - Cost estimating.
   - Engineering/technical.
   - Industrial engineering.
   - Management information systems.
   - Manufacturing/production.
   - Marketing and sales.
   - Materials management.
   - Packaging engineering.
   - Product servicing.
   - Purchasing.
   - Quality/reliability.
   - Subcontractors, as necessary.
   - Tooling engineering and maintenance.

2. A company must address nonconformities, whether internal or external, with disciplined problem-solving methods. They include:
   - For Chrysler suppliers, use of corrective action plans following the Chrysler Seven Disciplines (7D) format. Details include:
     – Description of the problem/defect.
     – Definition/cause.
     – Interim action and effective date.
     – Permanent action and effective date.
     – Verification.
     – Control.
     – Prevention.
   - For GM suppliers, use of the documented *Problem Reporting and Resolution Procedure*.

3. A company must attack actual or potential nonconformities by using mistake-proof methodology during problem resolution as well as during the planning of:
   - Equipment.
   - Facilities.
   - Processes.
   - Tooling.
4. A company must show that it understands and uses appropriate methods for continuous improvement, including:
   - Analysis of motion/ergonomics.
   - Benchmarking.
   - Capability indexes.
   - Control charts.
   - Cost of quality.
   - Cumulative sum charting.
   - Design of experiments.
   - Evolutionary operation of processes.
   - Overall equipment effectiveness.
   - Parts-per-million analysis.
   - Problem solving.
   - Theory of constraints.
   - Value analysis.

TECHNICAL GUIDELINES

ISO 9000 provides general guidance on quality improvement methods. ISO 9004-4: *Quality Management and Quality System Elements—Part 4: Guidelines for Quality Improvement* is a brief compendium of very valuable information on effecting continuous quality improvement in any kind of process. (These are *guidelines* only.)

The document begins with a set of definitions, including two not found elsewhere (i.e., in ISO 8402): *supply chain* and *correction*. The document goes on to discuss the following topics and subtopics:

- Fundamental concepts:
  - Principles of quality improvement.
  - Environment for quality improvement.
  - Quality losses.

- Managing for quality improvement:
  - Organizing for quality improvement.
  - Planning for quality improvement.
  - Measuring quality improvement.
  - Reviewing quality improvement activities.
- Methodology for quality improvement:
  - Improving the whole organization.
  - Initiating quality improvement projects or activities.
  - Investigating possible causes.
  - Establishing cause-and-effect relationships.
  - Taking preventive or corrective actions.
  - Confirming the improvement.
  - Sustaining the gains.
  - Continuing the improvement.
- Supporting tools and techniques:
  - Tools for numerical data.
  - Tools for nonnumerical data.
  - Training in applying tools and techniques.

Appendix B presents information on these most commonly used tools for analyzing numerical and nonnumerical data:

- Data collection form.
- Affinity diagram.
- Benchmarking.
- Brainstorming.
- Cause-and-effect (fishbone) diagram.
- Control chart.
- Flow chart.
- Histogram.
- Pareto diagram.
- Scatter diagram.
- Tree diagram.

Descriptions and examples, as well as details on application and procedures, are given.

PART THREE
Who's Who and
What's What in QS-9000

18. What is the Automotive
Industry Action Group (AIAG)?

The AIAG is an automotive trade association set up to improve communication and consistency among North American automotive manufacturers, suppliers, and customers.

AIAG task forces explore opportunities for what it calls "commonization" of business practices and processes. Some of AIAG's efforts are specific to certain sectors within the community (i.e., seating makers, exporters). Some of AIAG's initiatives have had more "global" reach, including Nafta issues, electronic data exchange, and QS-9000.

AIAG has developed and published guides that standardize key automotive quality practices. These guides, which are incorporated by reference into the QS-9000 standard, include:

- Advanced Quality Planning and Control Plan (APQPCP) (Question 30).
- Basic Statistical Process Control (SPC) (Question 68).
- Failure Mode and Effects Analysis (FMEA) (Question 30).
- Measurement Systems Analysis (MSAM) (Question 64).
- Production Part Approval Process (PPAP) (Question 39).

> **CAPSULE ANSWER**
>
> AIAG is a trade association set up to help automotive industry customers and suppliers harmonize, commonize, and communicate better.

AIAG is also the publisher of the various documents in the QS-9000 system (Question 15).

Like most trade associations, AIAG, a not-for-profit group, sponsors trade shows, provides education, and offers an extensive library of publications that, as it says: "involve virtually every technological area in the automotive industry."

19. What is the ISO?

"ISO" (usually pronounced "ice-oh") is not a true acronym. It is a universal nickname for the International Organization for Standardization. "Iso," the Greek root for the word "equal" (isometric, isosceles, and so on), fits the organization. ISO is one of the world's largest organizations involved in creating and publishing international standards to promote world trade.

> **CAPSULE ANSWER**
>
> ISO, the creator of ISO 9000, is the International Organization for Standardization, an international body dedicated to creating voluntary standards to promote global trade.

Formed in 1947 and based in Geneva, Switzerland, ISO counts some 116 nations as member bodies (actively involved in the technical committees and other activities of the organization), correspondent members, and subscriber members.

ISO's stated objectives are:

- To promote development of standardization in order to facilitate international exchange of goods and services.
- To promote cooperation in intellectual, scientific, technological, and economic activity.

The chief product of ISO is a body of international agreements that are then published as voluntary international standards. The volume of this work is quite impressive. For example, ISO published 66 new and revised standards between August and September of 1993. Standards cover nearly every field of commercial activity except electrical and electronic engineering, which are dealt with by a separate body, the International Electrotechnical Commission (IEC). Together, ISO and IEC comprise the largest nongovernmental system for voluntary industrial and technical collaboration.

Most of ISO's member groups are national standards bodies incorporated under the public laws of their respective countries. The rest—including the U.S. representative, ANSI (American National Standards Institute)—are nongovernmental organizations. Member groups are organized into several thousand "technical committees," each of which is responsible for a particular field of standards. For example, there are technical committees on welding (TC/44), essential oils (TC/54), small craft (TC/188), and sieves (TC/24). The technical committee responsible for ISO 9000 is TC/176.

The creation of an international standard begins with discussions among the members of a technical committee. From these discussions comes a *committee draft,* which is circulated among committee members for analysis and comment. When the committee reaches consensus on the draft, it is published by ISO as a *draft international standard* and is submitted to all ISO member bodies for voting. Publication as an international standard requires the approval of at least 75 percent of the member bodies casting votes.

ISO has no official relationship with the Big 3 or the Automotive Industry Action Group (AIAG) (Question 18). It is not involved in the development of QS-9000. However, it generally supports the development of industry-specific standards such as QS-9000. ISO 9000 has always been intended to serve as a "basic" or "floor" quality system standard on which individual industries and suppliers/customers can structure their own more specific requirements, as is the case with QS-9000.

ISO's role is limited to the development, publication, and revision of standards. It does not enforce, regulate, or audit—nor does it publish, or plan to publish, quality standards more specific to particular products or services.

Beyond the requirements documents (ISO 9001, 9002, and 9003), ISO does provide broad-based interpretive guidance. For example, one series of documents covers:

- Application of the standards (generally).
- Application of the standards to general product/service categories, including:
  - Services.
  - Processed materials.
  - Software.
- Auditing.

- Configuration management.
- Dependability management.
- Management of measuring equipment.
- Quality improvement.
- Quality planning.

These documents are not part of QS-9000 and are not even referenced by QS-9000. But they can be very helpful in the interpretation and application of the ISO 9000 portions of QS-9000.

20. What is an accreditation body?

Odds are that a company will never deal directly with an accreditation body. The company will focus on choosing a registrar, which then deals with an accreditation body. Choice of a registrar is partly limited by QS-9000 rules (Question 91).

An accreditation body is to a registrar what a registrar is to a company.

> **CAPSULE ANSWER**
>
> Accreditation bodies audit registrars, make sure they comply with Big 3 and international "rules," and issue certificates to them—just as registrars issue certificates to qualifying companies.

The accreditation body evaluates the registrar—by audits and by other means—against a defined and documented standard. When the accreditation body finds that the registrar meets the requirements, the accreditation body "accredits" the registrar. A registrar puts its credibility on the line to register a company's system as compliant with QS-9000. An accreditation body puts its credibility on the line when it declares that a registrar is capable, competent, and compliant with international rules.

Usually, accreditation bodies are government agencies, or organizations sanctioned by their national governments. Typically, an accreditation body is at the top tier of the quality system chain, just below the national government agency that accredits it, and just above any number of registrars that it accredits.

For example, the accreditation body in the United Kingdom is the United Kingdom Accreditation Service (UKAS, once known as NACCB). It is sponsored by the British government's Department of Trade and Industry, which is equivalent to the U.S. Department of Commerce. Registrars accredited by UKAS are permitted to display the UKAS logo. Registrations issued by these registrars carry the UKAS logo as well. Similar accreditation bodies operate in Australia, Belgium, Canada, Germany, the Netherlands, and several other countries.

In the United States, the generally recognized accreditation body is the Registrar Accreditation Board (RAB). This group does not have U.S. government sanction. It is a joint venture sponsored by two private industry groups, the American Society for Quality Control and the American National Standards Institute. RAB has accredited a number of registrars, from North America and elsewhere.

The Big 3 recognize only accreditation bodies that comply with published "Accreditation Body Implementation Requirements." These include:

- Requirements for the accreditation bodies themselves, such as:
  - The number of audits they must witness before accrediting a registrar.
  - Definition of de-listing criteria.
  - Maintenance of a listing of accredited registrars.
  - Form of accreditation certificate.
- Requirements that the accreditation body must impose on registrars, such as:
  - Compliance with the "Code of Practice," which includes compliance with a European standard, EN 45012.
  - Maintenance of a list of QS-9000 qualified assessors.
  - Qualifications of its board of directors.
  - Composition and qualifications of assessors and audit teams.
  - Rules for updating ISO 9000 certificates to QS-9000.
  - De-listing criteria.

QS-9000 candidates' choice of registrars is pretty much limited (Question 91). Only accredited registrars are acceptable, which is a good thing. Registrars that are not accredited often seek to cash in on ISO 9000 candidates by selling certificates that have a distinct lack of credibility.

Even though a QS-9000 candidate's choice of registrars is limited, beyond North America not all registrars are accredited by all accreditation bodies. A company that does business elsewhere in the world, or plans to, will find it worthwhile to select a registrar that is accredited not only by RAB (the U.S. group), but also by one of the overseas accreditation bodies.

21. What is a quality systems registrar?

A registrar, or "registration body" (the most preferred term), is sometimes called a "certification body" or even an "accreditation body." (Accreditation bodies are entirely different; see Question 20.) The registrar is the organization that checks a quality system and states whether it meets QS-9000 requirements for a prescribed and agreed-on period of time.

> **CAPSULE ANSWER**
>
> A registrar audits quality systems, registers conforming quality systems to QS-9000, and oversees continued conformance to the Standard. The registrar is bound by two sets of fairly strict rules.

The registrar's duties are:

- Audit a company's quality system to determine the degree of conformity to QS-9000 standards. The audit is carried out on paper (in a desktop study) and on site (throughout the facility).
- Register the quality system, assuming it conforms, to QS-9000.
- Monitor the conformity on an ongoing basis by means of regular follow-up audits and other methods.

All quality system registrars perform these functions within fairly uniform protocols, but they are not all the same. Registrars differ in two principal ways: their accreditation status, and the scope of their accreditation.

As discussed in Question 20, QS-9000 registrars must be accredited by a Big 3-approved accreditation body. An accredited registrar is bound by two sets of rules. The first set, handed down by the Big 3, mandates the following:

1. Compliance with the QS-9000 "Code of Practice," which imposes the following:
   - Accreditation by a Big 3-recognized accreditation body.
   - Adherence to a European standard, EN 45012.
   - Restrictions on registrars that have provided quality system consulting services (including on-site training services) to the client prior to August 1994.
   - Assessment of suppliers:
     - By assessors that have completed AIAG-sponsored training courses.

- To QS-9000 requirements, including all elements of suppliers' quality systems (as stated in *Quality System Assessment*) implemented to meet automotive customer needs.
- To check for effective implementation of requirements as well as ongoing practices.
- To review the handling of customer complaints; internal audit (Question 32) and management review (Question 31) results and actions; and progress toward continuous improvement (Question 27) targets.
- To confirm that quality system consultants are not included, except in the role of observers.
■ Repetition over the entire quality system a minimum of once every three years.
■ A fully documented report that includes opportunities for improvement.

2. Maintenance of a listing of QS-9000 qualified assessors who have passed sector-specific training courses (provided by AIAG).
3. Representation on the board of directors by persons who have automotive industry experience.
4. Audit teams with at least one member who has relevant auto industry experience.

The other set of rules is older, international, and quite strict. It is a European standard published in a document referred to as EN 45012: "European Standard for Bodies Certificating Suppliers' Quality Systems." These guidelines, which have the force of rules, govern the organization and activities of registrars:

1. Registrars must make their services available to all qualified suppliers without imposing "undue" financial or other conditions, and must administer their regulations in a nondiscriminatory manner.
2. The registrar's organization must not engage in activities that may affect its impartiality. For example:
   ■ It must not provide consulting services "on matters to which its certificates are related" (i.e., quality systems). This requirement is superseded by the QS-9000 restriction noted above.
   ■ It must not directly engage in commerce with firms that it has assessed and/or registered.

- Individuals involved in the registration process must not have provided consulting services to registration clients, or any related firms, within the previous two years (after August 1994).
- Its employees and agents must not engage in business activities that would cause others to question the firm's impartiality.
- It may not market consultancy and registration services together, and may not recommend consulting services to clients.
- Its auditors may not give advice as part of registration audits.
- It must provide the accreditation body with documentation of its employees' qualifications.
- It must have appropriate facilities for carrying out its activities.
- It must have a quality manual and documented procedures. (Curiously, neither the Big 3 nor EN 45012 require that registrars register to ISO 9000!)
- It may not grant, or renew, certificates of registration until all major noncompliances are eliminated.

Another point of differentiation is the scope of the registrar's accreditation. All registrars are not accredited, or approved, to register firms in any line of business. Each registrar is accredited to operate within the business or industrial sectors about which it has documented expertise (generically referred to as the registrar's scope).

For information on other ways in which registrars vary, and guidance on how to select the best registrar, see Question 91.

PART FOUR

Technical Requirements and Guidelines

22. Who is responsible for the facility's quality policy and objectives?

Top management. No exceptions; this is a must. Company management, especially "the highest level of management," is responsible for determining the company's quality policy, writing it down, and making sure everyone understands it on an ongoing basis, not just at the start.

> **CAPSULE ANSWER**
>
> Top management must determine and write down the company's quality policy, objectives, and commitment. It must also make sure that everyone in the company understands the policy and works in a way that is consistent with it.

The quality policy should not be a "puff piece" full of warm fuzzy phrases that sound great but mean little. The quality policy, the Standard asserts, must include quality objectives that are "relevant to . . . organizational goals and the expectations and needs of customers." The latter phrasing of that statement is especially important. It links the quality system to the activity that every company must carry out in order to survive: satisfying the customer.

Having written the quality policy down, management must demonstrate its commitment to it by running the business in a way that is consistent with the policy and is aimed at fulfilling the stated quality objectives.

A meaningful, relevant quality policy is one that includes solid goals, is championed by management, and is understood and worked to by all employees. Such a policy acquits the QS-9000 system of the charge, leveled by its detractors, that it is a "meaningless paperwork exercise." Having and adhering to such a policy is the very first requirement in the Standard. And, arguably, it is the requirement that is least effectively met.

TECHNICAL REQUIREMENTS

Management must:

1. Define and document the facility's quality policy, including quality objectives and quality commitment.
2. Ensure that the policy is understood, implemented, and maintained at all levels of the organization, and that it is relevant to the company's corporate goals and the expectations and needs of its customers.

TECHNICAL GUIDELINES

Management should:

- Develop a quality policy that takes into account:
  - The grade of service to be provided.
  - The organization's image and reputation for quality.
  - The objectives for service quality.
  - The approach to be adopted to achieve the quality objectives.
  - The role of company personnel who are responsible for implementing the policy.
- Establish a policy for service quality and customer satisfaction.
- Document the objectives and commitment pertaining to key elements of quality, such as fitness for use, performance, and so on.
- Document specific quality objectives that are consistent with the quality policy and the other objectives of the organization.
- Ensure that the quality policy is:
  - Expressed in easy-to-understand language.
  - Relevant to the organization's other policies and products.
  - Ambitious and achievable.

- Take all necessary measures to ensure that the quality policy is understood, implemented, reviewed, maintained, and promulgated at all levels of the organization.
- Demonstrate the quality commitment visibly, actively, and continually.
- Follow up on implementation of the quality policy.
- Sanction or reject deviations or exceptions to the quality policy.

REQUIRED DOCUMENTS

1. The complete quality policy statement, circulated to all employees as a quality manual, and posted on signs and other media on the premises.
2. A "capsule version" for printing on the back of business cards and on pocket cards distributed to employees.

AUDIT ISSUES

Assessors will ask employees at random about the quality policy. It is not necessary or even desirable that employees be able to recite it from memory. An ideal response occurs when an employee explains how the policy affects his or her own job.

23. What responsibilities does the Standard delegate to company management with respect to quality?

In essence, QS-9000 requires management to manage; to lead; to structure the business, equip it with the needed resources, operate it in accordance with a written plan (Question 24), and analyze results objectively, especially with

> ### CAPSULE ANSWER
>
> Management is required to: define the authority and interrelation of people who affect quality in any way; assign employees who are competent to perform their tasks; and identify and provide needed resources.

respect to the all-important goal of customer satisfaction. If this seems like "Business 101" stuff, it *is:* every company does these things to varying degrees and with varying levels of effectiveness. But most companies do not do all of them on a disciplined and documented basis.

TECHNICAL REQUIREMENTS

Management must fulfill these activities:

1. Define the responsibility, authority, and interrelation of personnel who manage, perform, and verify work affecting quality. This applies especially to people who need freedom and authority to identify or prevent problems, find solutions, or otherwise control nonconforming product.
2. Operate key aspects of the business—concept development, prototype, and production—in a way that involves all the contributing functions in the decision-making process.
3. Identify resource requirements, provide adequate resources, and assign trained personnel for management, performance of work, and verification activities.
4. Assign a member of management (usually called a "management representative"; Question 75) with the authority to:
   - Oversee the establishment, implementation, and maintenance of the quality system in accordance with the Standard.
   - Report to management on the performance of the quality system.
   - Liaise with external bodies on matters concerning the quality system.
5. Utilize a written business plan (Question 24).

TECHNICAL GUIDELINES

Management should:

- Define the quality factors affecting market position and the objectives relating to new products, processes, or services, in order to allocate resources on a planned basis.
- Create an appropriate image for the company, based on real actions taken to meet customer needs.
- Emphasize identification of actual or potential quality problems and initiation of preventive measures.
- Delegate responsibility and authority clearly to each activity, so that assigned quality objectives can be attained with desired efficiency.
- Define interface control and coordination among activities.

- Establish a clear organizational structure for quality within the overall management of the organization, with well-defined lines of communication.
- Encourage effective interaction between customers and service organization personnel.
- Provide appropriate and sufficient resources to achieve quality objectives that require:
  - Human resources.
  - Equipment for design, manufacturing, inspection, and tests.
  - Computer software.
- Provide the equipment and supplies necessary to provide a product and/or service, including:
  - Quality assessment facilities.
  - Instrumentation.
  - Computer software.
  - Operational/technical information.
- Establish a comprehensive plan for audits of all aspects and components of the quality system, to determine whether various elements are effective in achieving stated quality objectives.
- Define all employee responsibilities, and explicitly define the responsibility and authority of all personnel whose activities influence quality.
- Determine the levels of competence, experience, and training that are necessary to ensure the desired capability of personnel.
- Ensure that the quality system promotes continuous quality improvement (Question 27). Suggested tactics are:
  - Encouraging supportive management styles.
  - Promoting values, attitudes, and behaviors that foster improvement.
  - Setting clear quality improvement goals.
  - Encouraging effective communication and teamwork.
  - Recognizing successes and achievements.
  - Training and educating for improvement.
- Ensure that the management representative designated to monitor and report on quality enjoys access to the highest levels of management in the organization. If the management representative has other duties, ensure that there is no conflict of interest.

Management bears a heavy load of responsibility, but the Standard does not exempt other employees. By understanding the scope, responsibility, and authority of their functions, and their impact on quality, all employees should feel responsible for quality.

Even customers ("purchasers") are not exempt. ISO 9000's guidelines list these obligations:

■ Provide all necessary information to the company.
■ Assign a representative who has the authority to discuss and resolve contractual matters with the company.
■ Define acceptance criteria and procedures.
■ Define purchasing requirements.
■ Answer questions, conclude agreements, and ensure that personnel employed by the purchaser observe the agreements.

REQUIRED DOCUMENTS

1. An organization chart that covers all functions down to the supervisory level. The chart should be dated, signed, controlled, and included, at a minimum, in the quality manual. Because the chart is a living document, it is best to include only functional titles (not personal names), to reduce the frequency of needed revisions. Every functional title mentioned in the quality manual should be included on the organization chart so that the "responsibility, authority, and interrelationship" can be understood.
2. Job descriptions. The qualifications of all functions that affect quality must be documented. By creating or adapting brief job descriptions (for all functions, including management), the requirements of Elements 4.1 and 4.18 of QS-9000 can be met (Question 29).

AUDIT ISSUES

Key members of management must understand the relationship between stated quality objectives and available resources: "What resources are provided to help the company meet its stated quality objectives?"

The requirement for cross-functional decision making is especially difficult for companies that have not made a conscious effort to do this previously. Auditors will be examining the quality and design planning processes, in particular, to determine the extent to which all key functions are involved.

24. Does QS-9000 really require a written business plan?

CAPSULE ANSWER

Management is required to define, document, and maintain a business plan that, at a minimum, ties in to customer needs and is subject to objective measurement.

Yes, and it is not kidding around, either. A company must be prepared to show that the business is run in accordance with a documented business plan—a living document that includes goals driven by customer needs. Further, it must be able to show an analysis of performance against the business plan and the stated goals, using objectively acquired information.

This is a point on which the Standard butts heads with a company's legitimate need to protect confidential and/or proprietary information. For that reason, the Standard states that the content of the business plan is not subject to third-party audit. Certain exemptions for safeguarding of information are also made for companies that compete directly with a customer division. Some assessors merely check to be sure a plan exists, but no candidate for registration should count on that treatment.

TECHNICAL REQUIREMENTS

1. The business plan must be formal, documented, and comprehensive.
2. It must tie in to methods for determining customer expectations.
3. Referenced data must be collected via an objective and valid process.
4. Methods to review, update, and communicate the plan to affected functions must be documented.
5. The business plan must be a controlled document.

TECHNICAL GUIDELINES

The plan should include (as applicable):

■ Financials, market and financial projections, plant and facility plans, cost objectives, quality objectives, and general objectives.
■ Short-term (1–2 years) and long-term (3 years plus) goals.
■ References to competitive products as well as internal and external benchmarking studies.

REQUIRED DOCUMENTS

1. Business plan (and associated control documents).
2. Procedure for creating, updating, and utilizing the business plan.

AUDIT ISSUES

Auditors will probe the business plan process to be sure it takes into account all legitimate inputs relevant to the business. They will also assess the extent to which relevant functions are involved in creating, updating, and/or utilizing the business plan. In short, assessors want to satisfy themselves that the document is truly used for strategic business planning and is not simply window dressing.

25. What is the goal of a QS-9000 quality system?

The goal of a QS-9000 quality system, as stated in *Quality System Requirements QS-9000*, is to "provide for continuous improvement, emphasizing defect prevention and the reduction of variation and waste in the supply chain."

CAPSULE ANSWER
The goal of a QS-9000 quality system is to foster continuous improvement, defect prevention, and reduction of waste and variation.

But a broader, more comprehensive goal is expressed succinctly in the ISO 9000 guidance documents: To ensure that the company's output conforms to requirements specified by the customer. These requirements are the customer's expression of needs and expectations, both stated and unstated, as:

- Determined by the marketing function.
- Translated into technical specifications.
- Converted into designs.
- Fulfilled, by the process, as products.

To ensure this conformity, the quality system must be planned, understood, implemented, maintained, and provable.

TECHNICAL REQUIREMENTS

1. A quality system must be structured to ensure that a product conforms to specified requirements.

TECHNICAL GUIDELINES

The quality system should accomplish the following:

APPLYING THE STANDARD

"Specified requirements" includes the stated and implied expectations of customers (taking into account the provisions of contract review and design control).

- Ensure that adequate and continuous control is exercised over all activities affecting quality.
- Provide confidence to management and employees that:
  - Quality requirements are fulfilled and quality improvement is taking place.
  - The system is well understood and effective.
  - Products and services actually do satisfy customer expectations.
  - Emphasis is on problem prevention rather than detection after occurrence.

REQUIRED DOCUMENTS

1. The quality manual, procedures, instructions, and any other documents of the quality system must be available.

26. What are the characteristics of a QS-9000 quality system?

A quality system is a documented, processwide operation, established and led by management, that conforms to the requirements of QS-9000. The QS-9000 documents are basically silent on specific characteristics, but ISO 9004-3,

CAPSULE ANSWER

A QS-9000 quality system is a documented, self-improving union of resources and activities that governs every aspect of a process that affects quality.

the ISO 9000 guidelines document for processed materials, provides a comprehensive description of the scope of a quality system and is quite relevant to QS-9000 as well: "The quality system typically applies to, and interacts with, all activities pertinent to the quality of a product, process, or service. It involves all phases from initial identification to final satisfaction of requirements and customer expectations."

These phases and activities may include the following:

- Marketing and market research.
- Technical research and development.
- Design/specification engineering and product development.
- Procurement.
- Process planning and development.
- Production process measurement, control, and adjustment.
- Production.
- Process maintenance.
- Inspection and testing.
- Packaging and storage.
- Sales and distribution.
- Customer use.
- Technical assistance.
- Disposal after use.

TECHNICAL REQUIREMENTS

1. The facility must establish and maintain a quality system that conforms to the Standard.
2. The quality system structure must be documented in a quality manual that reflects the Standard and includes or references the procedures relating to the details of the system.

TECHNICAL GUIDELINES

The quality system should:

- Be an integrated process throughout the entire life cycle of the product. It should ensure that quality is built in as development progresses, rather than being discovered at the end of the process (through inspection, for example).

- Be organized so as to control all technical, administrative, and human factors affecting the quality of products.
- Take into account and document with objective evidence in the form of information and data:
  - The company's interest in and need to attain and maintain desired quality at optimum cost via planned and efficient deployment of technological, human, and material resources.
  - Customers' needs and expectations, especially confidence in the supplier's ability to deliver, on a consistent basis, the desired level of quality.
- Include operational procedures that coordinate quality activities and specify objectives and required performance of various activities having impact on quality (Question 36).
- Indicate the establishment, within the overall organizational structure, of functions relating to the quality system, with lines of authority and communication clearly defined.
- Include quality control as an integral part of the service processes: marketing, design, and service delivery.
- Be structured and adapted to the facility's particular type of business.
- Take into account appropriate elements outlined in the Standard.
- Emphasize preventive actions that avoid occurrence of problems, without compromising the ability to respond to and correct failures that may occur.
- Use marketplace input to upgrade new and existing products and to improve quality system.

Feedback systems should be established among the interacting elements of the service process: marketing, design, and delivery. Feedback on quality system effectiveness should be gathered at points of service delivery from supplier assessments and customer assessments, as well as from internal and external quality audits.

ADVICE

The scope of a quality system encompasses all functions and requires involvement, commitment, and effective interworking of all personnel in the organization to achieve continuous improvement.

27. How is continuous improvement addressed in the Standard?

CAPSULE ANSWER

Continuous improvement is central to the effectiveness of an ISO 9000 quality system.

Continuous improvement is important enough to get its very own section in QS-9000—something it does not rate with ISO 9000. If rigorously assessed, this QS-9000 requirement is one of the toughest to meet. But the process that is required, if conscientiously followed, is also one of the things that makes QS-9000 more than just a "paperwork exercise." It can truly improve a company's ability to meet customer needs and thereby improve overall company performance.

TECHNICAL REQUIREMENTS

1. The company must deploy a continuous improvement system that results in demonstrated improvements in quality, service, and price.
2. The company must identify the processes that are most important to customers, and then develop action plans for continuously improving them. The stated goal is assurance that requirements for those processes are always met.
3. The improvement projects must be tracked via objective measurements so that actual improvement can be assessed.
4. The company must demonstrate knowledge of standard continuous improvement methods, as appropriate.

TECHNICAL GUIDELINES

Examples of potential areas for continuous improvement include:

- Customer dissatisfaction.
- Difficulty in assembly or installation.
- Efficient use of floor space.
- Excessive cycle time.
- Excessive handling and storage.
- First-run capability.
- Machine changeover.

- Measurement systems capability.
- Optimization.
- Quality costs.
- Scrap, rework, and repair.
- Testing requirements.
- Unscheduled machine downtime.
- Variation.
- Waste of labor.

ISO 9000 does not call out continuous improvement specifically, but its guidance documents offer extensive advice that can help companies meet the QS-9000 requirement. The ISO 9000 guidance documents suggest that management should:

- Establish an information system for the collection and dissemination of data from all relevant sources. Given that information, management should conduct continual evaluations of the operations involved in the process, to identify and actively pursue opportunities for quality improvement.
- Assign responsibilities for the information system, and for quality improvement, to specific individuals.

The continuous improvement program should be comprehensive. ISO 9000 says that the program should identify:

- Characteristics that, if improved, would most benefit the customer and the organization.
- Any changing market needs that are likely to affect the product or service to be provided.
- Deviations from specified quality due to ineffective or insufficient quality system controls.
- Opportunities for reducing costs while maintaining and improving the quality of product or service provided.

Required Documents

1. A procedure for identifying continuous improvement projects, carrying them out, and tracking the results.
2. Documented action plans or projects with measurables.

AUDIT ISSUES

For companies that have never focused specifically on continuous improvement, this requirement has audit targets galore. Here, auditors probe the system to make sure that the process for selecting continuous improvement projects is sound (are projects focusing on processes most important to customers?). They check for documentation of the projects, and they review the measurables to see how effective the system is.

28. Does the Standard require firms to set any formal qualifications for employees?

> **CAPSULE ANSWER**
>
> Management is required to set formal qualifications for employees in terms of education, training, and/or experience.

Yes. For all functions that affect quality, management is required to define and document "responsibility and authority." The Standard states elsewhere (clause 4.18) that "personnel affecting specific assigned tasks shall be qualified on the basis of appropriate education, training, and/or experience."

Some interpret this as requiring job descriptions. They may in fact fulfill this requirement, but job descriptions per se are *not* required. Most especially not required—and to be avoided—are job descriptions that are lengthy, inaccessible, time-consuming, non-value-added marvels of legalese.

The bottom line is best expressed in an ISO 9000 guidance document: Management must "determine the level of competence, experience, and training necessary to ensure capability of personnel." Figure that out, write it down, approve it and maintain it, and the formal qualifications are done.

TECHNICAL REQUIREMENTS

1. Personnel performing specified assigned tasks must be qualified on the basis of education, training, and/or experience, as appropriate. Although particularly important for employees performing tasks that affect quality, this requirement actually applies to all types of tasks in the organization.

2. The facility must maintain records proving that it sets and enforces appropriate qualifications for all employees, as follows:

- Where appropriate, employees should be certified in the skills needed to carry out their jobs. This is particularly important for functions for which independent, documented certification is available.
- Qualifications should especially be established and documented for those performing safety-related work.
- The facility should establish a system for periodic assessment of skills and/or capabilities. This may, as appropriate, include actual demonstrations of skills. All assessments should be documented.

29. Is training a requirement of the Standard?

Training is not just a requirement. QS-9000 strongly suggests that training must be "viewed as a strategic issue affecting all" employees.

In general, companies are required to define and document the means by which:

> **CAPSULE ANSWER**
>
> Management must implement systems for identifying training needs, meeting those needs, assessing their effectiveness, and keeping appropriate records.

- Training needs are identified.
- Training is carried out.
- Training is recorded.
- Effectiveness is assessed.

Incidentally, this applies not just to permanent full-time employees, but also to part-timers and temporaries. It also applies (most definitely) to "on-the-job training," which is still wholly permissible under a QS-9000 system but, as a rule, must be much more structured and formalized than it has typically been in the past.

One other note: The fact that a company lacks a dedicated human resources function or department should not keep it from fulfilling this requirement.

TECHNICAL REQUIREMENTS

1. Management must establish and maintain documented procedures for identifying training needs.

2. Management must provide training for all personnel performing activities that affect quality.
3. The company must maintain adequate records documenting its compliance with training requirements.
4. Training effectiveness must be assessed periodically.

TECHNICAL GUIDELINES

ISO 9000 guidance documents strongly recommend that employees be trained not only in the operation of the quality system, but also in the reasons for its implementation. Particular attention should focus on new personnel and on people transferred to new assignments, but consideration should also be given to periodic refresher courses for longer-term employees.

- Training can be carried out either internally or by an outside body.
- As a "strategic issue affecting all personnel," training should be driven from the top, even if specific training activities are developed and carried out locally. Training should also be a prominent component of various activities in the system, including:
  - Business planning (Question 24).
  - Advance quality planning (Question 30).
  - Management review (Question 31).
- Executive and management personnel should be provided with the following kinds of training:
  - Understanding of the quality system.
  - Tools and techniques needed for full executive management participation in operation of the system.
  - Criteria for evaluating system effectiveness.
  - Analysis of quality costs.
- Supervisors and line employees should be "thoroughly trained in methods and skills required to perform their tasks." Options include:
  - Proper operation of instruments, tools, and test equipment they use; reading of documentation they must understand.
  - The relationship of their duties to quality.
  - Safety in the workplace.
  - Basic statistical techniques, as well as process control, data collection and analysis, problem solving, corrective action, team working, communication skills.
  - Customer satisfaction.

- Technical personnel should receive training to enhance their contribution to the success of the quality system.
- Employees should understand basic statistical concepts (Question 68).

REQUIRED DOCUMENTS

1. A procedure covering training planning, training, and assessment of effectiveness.
2. A training plan.
3. Records of training and effectiveness.

AUDIT ISSUES

Training can create a feeding frenzy among assessors. They will check a system up, down, and sideways, by means such as these:

- Reviewing training records and then asking the employees what training they have had.
- Asking employees about their training, and then cross-checking their responses against the actual records.
- Challenging the credentials of people doing the training.

Every employee will require some type of training—at a bare minimum, some training in the quality system—if this requirement is to be fully met. No one should be overlooked.

Developing the Quality System

30. What does the Standard mean by "advanced quality planning"?

QS-9000 requires a documented system for quality planning, whether a company is "design-responsible" or not. The Standard references a separate manual (*Advanced Product Quality Planning and Control Plan Reference Manual*) and requires that it be "utilized" by candidate companies.

> **CAPSULE ANSWER**
>
> QS-9000 requires a fully documented, cross-functional, and comprehensive advanced product quality planning process to control and improve the quality of new and existing products.

The output of advanced quality planning is the *control plan,* a living document that tracks product production and facilitates improvement. To get to that advanced level, several activities are required:

- Definition of "special characteristics" and identification by their special symbols (Question 49).
- Use of cross-functional teams (Question 17).
- Feasibility reviews (see below).
- Process Failure Mode and Effects Analyses (Process FMEAs).
- Development of facilities, processes, and equipment (Question 51).
- Production Part Approval Process (PPAP) (Question 39).

ISO 9000 requires advanced quality planning in a more general sense; QS-9000 requires it in a very specific and prescriptive way. Companies that have been working with Big 3 customers for years are well accustomed to the quality planning process, control plans, FMEAs, PPAPs, and the rest.

Companies new to this are in for a lot of work that they have most likely not had to do—at least in a formalized way—in order to meet their customers' requirements.

TECHNICAL REQUIREMENTS

1. Utilize the *Advanced Product Quality Planning and Control Plan Reference Manual.*

2. Identify special characteristics and indicate them by the customers' own unique symbols (Question 49).

3. Identify and acquire the controls, processes, and skills needed to achieve the required quality.

4. Ensure the compatibility of design, production, inspection/testing, installation, servicing, and documentation.

5. Develop quality control, inspection, and testing techniques, including development of new instruments, as needed.

6. Identify, in sufficient time for the needed process capability to be developed, the requirements for process capability that exceeds the known state of the art.

7. Identify suitable verification activities at appropriate stages in product realization.

8. Clarify standards of acceptability for all features and requirements.

9. Identify and prepare quality records.

10. Use cross-functional teams during the advanced product-quality planning process for new or changed products.

11. Use techniques in the *Advanced Product Quality Planning and Control Plan Reference Manual*, or similar techniques.

12. Include the following functions, as appropriate:
    ■ Design.
    ■ Manufacturing engineering.
    ■ Outside suppliers.
    ■ Production.
    ■ Quality.

13. Target team activities to:
    ■ Develop and review control plans (Question 55).
    ■ Develop and review FMEAs (see below).
    ■ Develop special characteristics (Question 49).
    ■ Develop ways of reducing potential failure modes.

14. Feasibility reviews: Using the format in the *Advanced Product Quality Planning and Control Plan Reference Manual*, manufacturing feasibility can be confirmed by assessing the suitability for production, taking into account:
    ■ Design.
    ■ Engineering requirements.
    ■ Material.
    ■ Process.
    ■ Required statistical process capability.
    ■ Specified volumes.

15. Process FMEAs: Develop Process Failure Mode and Effects Analyses (FMEAs) for all "special characteristics" (Question 49). An FMEA:
   - Identifies the ways a process could potentially fail to meet the process requirements or design intent.
   - Identifies, classifies, and ranks the potential effect of such a failure.
   - Calls out potential causes, probabilities, and frequencies of such a failure.
   - Specifies the process controls in place to prevent or detect such a failure, and predicts the effectiveness of such measures.

An interim result is a *risk priority number* that enables ranking potential failure modes so that corrective action can be applied against those that are most serious or potentially damaging. Corrective action and responsibility are then assigned.

TECHNICAL GUIDELINES

Quality plan contents (included or by reference) are:

- Quality objectives (in measurable terms wherever possible).
- Specific allocation of responsibility and authority during different phases.
- Specific procedures, methods, and work instructions to be applied.
- Defined input/output criteria for each development phase.
- Identification of types of testing, verification, and validation activities to be carried out, including:
  - Schedules.
  - Resources.
  - Approval authorities.
- Methods for modifying and changing the quality plan as a project proceeds.
- Specific responsibilities for quality activities.

A quality plan use cycle has the following components:

- Preparation.
- Formal review by all organizations concerned with its implementation.
- Agreement.

- Assurance that it is understood and observed by the organizations concerned (through management reviews, internal quality audits, and other means).
- Updates, as needed.

A company should refer to the *Advanced Product Quality Planning and Control Plan Reference Manual* for full details, samples, and so on.

REQUIRED DOCUMENTS

1. Advanced product quality planning procedure.
2. Related documents, as specified above.

31. How important are the management reviews mentioned in the Standard?

Vital—and not just because they are required. They serve a vital purpose: keeping management focused on the quality system.

> **CAPSULE ANSWER**
>
> Management reviews, required by the Standard, are management's independent and comprehensive look at the results and effectiveness of the quality system.

One of the "knocks" on QS-9000 and ISO 9000 is that they are "flavor of the month" quality programs—here today, gone tomorrow. But the programs themselves don't cause "flavor of the month." Management does, by allowing itself to get distracted. After being hot on a program or a system for a quarter or two, its attention and enthusiasm get diverted to something else. "Flavor of the month" can only happen when management lets it happen.

The management reviews required by QS-9000 help prevent this fickleness. They require management to focus on the quality system on a scheduled basis, in a formal setting, and in a documented way.

When managers have gone through a few of these reviews, they make some pleasant discoveries. Management reviews can be excellent management, communication, and intelligence-gathering venues. And they reinforce the effectiveness of the QS-9000 system—which, in turn, helps boost the performance of the business.

TECHNICAL REQUIREMENTS

1. Management must review the quality system at defined intervals.
2. The review must include all the elements of the quality system (Elements 4.1 through 4.20; Part II and Part III, as relevant).
3. Management must assess the suitability and effectiveness of the quality system, that is, whether it meets:
   - ■ The requirements of the Standard.
   - ■ The company's quality policy.
   - ■ The company's quality objectives.
4. Records must be kept.

TECHNICAL GUIDELINES

For management reviews:

■ No specific interval is prescribed. The interval must, however, be "defined"—that is, specified in documented procedures. Annual reviews are suggested, once a system is implemented, registered, and at steady state. During implementation, reviews should be held more often—perhaps monthly.

> **APPLYING THE STANDARD**
>
> When specifying this and other intervals, don't box yourself in by being too specific. It's best to say "approximately twice per year" rather than "every six months."

■ Objectives might include:
   - – To increase the effectiveness and efficiency of the quality system.
   - – To seek opportunities for improvement.
   - – To consider updating the quality system in reaction to changes brought about by new technology, quality concepts, market strategies, or social/environmental conditions.
■ All "Relevant sources of information" should be considered. Suggested areas for review include:
   - – Organizational structure.
   - – Adequacy of staff and resources.
   - – Comparison of achieved quality with required quality.
   - – Feedback on performance of process and product.
   - – Specific chronic problem areas.
   - – Results from internal audits.

■ All observations, conclusions, and recommendations generated should be "submitted in documentary form" to top management for necessary action.

REQUIRED DOCUMENTS

1. A procedure for carrying out management reviews.
2. Records of management reviews.

AUDIT ISSUES

A management review must address ALL elements of the quality system, and this coverage must be documented in the minutes.

32. What are internal quality audits all about?

First, let's talk about what internal quality audits are *not* about. They are *not* about policing the system, writing people up, or scoring points off others.

Internal quality audits are fair and objective assessments of the entire qual-

> **CAPSULE ANSWER**
>
> Internal audits are carried out by trained, independent employees, to help implement the quality system and to improve it on an ongoing basis.

ity system. They are carried out on a scheduled basis, following a written procedure, and their results are documented. Internal auditors must have appropriate training, and they must be independent of the direct management of the area they are auditing.

There are two stated purposes for internal quality audits:

1. To see how well the activities in the workplace comply with the documented system.
2. To assess the effectiveness of the system.

During implementation, the first purpose is more important; internal audits are actually catalysts for compliance (Question 88).

In addition, a properly selected, well-trained audit team helps carry the quality system message to the rest of the organization. For this reason, it is essential that the audit team be selected from as wide a range of levels and functions as possible—not just managers; not just office people.

After initial implementation and as the system matures, compliance becomes less of an issue and assessing effectiveness become more important. Many companies miss this point. Their internal audits stay focused on compliance and become rote activities. Auditing for effectiveness requires a more probing, intuitive, and analytical audit technique.

TECHNICAL REQUIREMENTS

1. Internal quality audits must be conducted on a scheduled basis in accordance with a procedure.
2. Audits must be carried out by trained people who are independent of the direct management of the audited area.
3. The frequency of audits depends on:
    - The importance of the audited area (the more critical to quality, the more frequent the audits).
    - Prior audit results (the more problems, the more frequent the audits).
4. Audits must assess the suitability of the working environment in the audited areas.
5. Audit results must be documented.
6. Audit results must be presented to the management of the audited area.
7. Management of audited areas must take timely corrective action against any deficiencies found during an audit.
8. The effectiveness of corrective actions taken must be verified and recorded.
9. Records must be kept of internal audit activities.
10. Chrysler suppliers must assess their entire quality system internally at least once each year, unless directed otherwise. (Once a year is a bare minimum in any company.)

TECHNICAL GUIDELINES

The purposes of internal quality audits are:

- To provide an objective evaluation of:
  - Administrative and operational procedures.
  - Documentation.

- Equipment.
- Items being produced (to assess the degree of conformity to standards and specifications).
- Material resources.
- Operations.
- Organizational structures.
- Personnel.
- Processes.
- Record keeping.
- Reports.
- Work areas.

■ To determine the conformity of the quality system with documented requirements.
■ To measure the effectiveness of the quality system in meeting specified quality objectives.
■ To meet regulatory requirements.
■ To provide opportunities for improvement.
■ To facilitate external quality assurance.
■ To assess and document the implementation and effectiveness of corrective actions resulting from previous audits.

Internal quality auditors:

■ Should represent as broad a range of levels, functions, and responsibilities as possible.
■ Should rotate in and out of the audit team over months and years.
■ Need NOT be "certified."
■ Need NOT be led by a "registered lead auditor."

The audit plan should be established by management (usually, the management representative manages the internal audit system; see Question 75) and should specify:

■ The specific areas and activities to be audited.
■ The qualifications of people carrying out the audits.
■ The basis or reason for the audit:
    - Routine check.
    - Chronic nonconformities.
    - Customer complaints.
    - Organizational changes.

- Procedures for reporting audit findings, conclusions, and recommendations.
- The audit schedule—a system should be assessed once per quarter during implementation, twice in the year following registration, and then (overall) once per year thereafter, adjusted as required by audit results.

As part of audit reporting and follow-up:

- Observations, conclusions, and agreements on timely corrective action should be recorded and submitted for appropriate action by the auditee's management.
- The audit report should cover:
  – All examples of nonconformities.
  – Specific examples of nonconformities.
  – Possible reasons, where evident.
- Management responsible for the audited activity should ensure that necessary and appropriate corrective actions are taken.
- The results of internal quality audits are always a major focus of management review meetings (Question 31).

For detailed guidance, see the ISO 10011 series on auditors, auditing, and the audit process. Note that this is a guidance document only.

REQUIRED DOCUMENTS

1. Procedures for internal quality auditing.
2. Audit schedule and plan (by department, procedure, and month).
3. Documentation of training of auditors.
4. Records of internal audit activities, including audit reports, corrective action requests, and follow-up activities.

AUDIT ISSUES

When auditors check a company's audit schedule and plan, they will expect to see that the frequency of audits has been reviewed and, if necessary, adjusted in light of audit results. The entire system should be assessed internally within a defined period of time—say, once per year.

Auditors will examine audit reports for evidence to support their findings. Even positive findings need to be supported by objective evidence. It is not enough to write "OK" repeatedly alongside audit checklist items. Because the effectiveness of corrective actions must almost always be verified, it is important to document exactly how verification is done.

33. Are the requirements for quality records really as tough as they sound?

> **CAPSULE ANSWER**
>
> QS-9000 mandates strict and prescriptive requirements for the gathering and retention of quality records in order to verify conformance to requirements and the effectiveness of the quality system.

QS-9000 has strict record-keeping requirements. In fact, only five sections from Parts I and II of the Standard do *not* mention record keeping.

This strictness is imposed partly because QS-9000 is a *provable, verifiable* quality system. To become registered (Question 90), a firm undergoes a quality system audit carried out by an objective third-party registration body. The purpose of the audit is to verify that the quality system described in the quality manual (and other documentation) is alive, well, and in actual use. Auditors do not rely on verbal assurances. They are leery of believing even what they see with their own eyes. Only well-managed records are "objective evidence" (proof) that the auditee, over time, has been operating a conforming quality system.

Another reason for QS-9000's strict record-keeping requirements is simple business sense. Records serve these invaluable purposes:

- Identifying the causes of problems.
- Settling disputes and misunderstandings.
- Documenting claims.
- Establishing baselines.

In point of fact, conformity to QS-9000 requires only an amount of record keeping that makes good business sense. And, the Standard allows wide latitude in how the records are created, organized, and stored. Control is mandated; the Standard does not specify how control is to be carried out.

Unlike ISO 9000, QS-9000 is rather prescriptive about record retention intervals, as indicated by the following requirements.

TECHNICAL REQUIREMENTS

1. "Establish and maintain documented procedures" for the following activities involving quality records:
   - Accessing.
   - Collating.
   - Disposing.
   - Filing.
   - Identifying.
   - Indexing.
   - Maintaining.
   - Storing.

2. The records to be maintained under this requirement—which can be hard copy, electronic, or in some other form—are those specifically designated in the relevant sections of QS-9000. They include, at a minimum, documents pertaining to the following quality system areas:
   - Assessment of subcontractors.
   - Contract review.
   - Control of inspection, measuring, and test equipment.
   - Corrective action.
   - Customer-supplied product.
   - Inspection and test records.
   - Inspection and test status.
   - Inspection and testing.
   - Management review.
   - Nonconformity review and disposition.
   - Product identification and traceability.
   - Quality records.
   - Quality system.
   - Special processes.
   - Training.
   - Urgent product release.

3. Records must be:
   - Legible.
   - Stored and retained so as to be readily retrievable.
   - Kept in an environment that ensures minimal deterioration, damage, or loss.
   - Made available for review by customers or their representatives, as contractually agreed, for a specified period.
   - Kept for documented retention period(s).

4. Retention periods always default to specific customer or governmental requirements, as relevant. Default minimums are as follows:
   - One calendar year, plus the length of time a part (or a parts family) is active for production or service requirements, for:
     - Production part approvals.
     - Purchase orders and amendments.
     - Tooling records.
   - One calendar year after the year created, for quality performance records, such as:
     - Control charts.
     - Inspection and test records.
   - Three years for records of:
     - Internal quality audits.
     - Management reviews.
   - Copies of documents related to superseded parts, needed for new-part qualification, must be kept in new-part file.

TECHNICAL GUIDELINES

WHAT TO SAVE

- Documents sufficient to follow and demonstrate conformity to specified requirements and effective operation of the quality system.
- Documents needed to identify trends in quality measures as well as the need for, and the effectiveness of, corrective action.
- Documents that provide information on:
  - Analysis to identify quality trends.
  - Competitive comparisons.
  - Corrective action and its effectiveness.
  - Degree of achievement of quality objectives.
  - Level of customer satisfaction.
  - Results of quality system reviews.
  - Skills and training of personnel.
- Documents pertinent to:
  - Audits.
  - Corrective actions.
  - Designs.
  - Inspections.
  - Related results.
  - Reviews.

- Surveys.
- Testing.
■ A few examples:
  - Audit reports.
  - Calibration data.
  - Design records.
  - Drawings.
  - Inspection procedures and instructions.
  - Inspection records.
  - Material review reports.
  - Procedures.
  - Qualification reports.
  - Quality cost reports.
  - Quality surveys, audits, and reviews.
  - Specifications.
  - Test procedures.
  - Test records.
  - Validation reports.
  - Work instructions.

CONDITION OF DOCUMENTS—GUIDELINES

Documents must be maintained with care, to keep them:

1. Clean.
2. Dated (including revision dates).
3. Legible.
4. Protected from unauthorized alteration.
5. Readily identifiable and retrievable.
6. Safely stored.

DISPOSAL OF DOCUMENTS—GUIDELINES

■ When setting policy, a facility should take into account:
  - Customer requirements (see above).
  - Regulatory authority.
  - Product liability/legality.
■ Documents should be removed and/or disposed of when outdated.

QUALITY RECORDS POLICY

A formal policy statement should spell out the procedures for:

- Changes and modifications of documents.
- Availability of records to customers and subcontractors.

REQUIRED DOCUMENTS

1. Quality records procedure.
2. Back-up procedure (or work instruction) for electronic records.
3. Quality records roster or log, identifying type, location, retention interval, and so on.

AUDIT ISSUES

Auditors will spot-check records to make sure that they are where they are supposed to be and that minimum retention requirements are being met. A company should avoid the wording, "Records are disposed of after three years." A better statement is: "After three years, records are subject to disposal." If an auditor finds records that are, say, a week past their disposal period, this wording protects against getting a noncompliance.

Accessibility is a big issue with some auditors. One even had his own test for accessibility: when he requested it, a record had to be produced within ten seconds!

Documenting the Quality System

34. What are the QS-9000's documentation requirements?

Documentation is the heart of QS-9000. In general, four different types of documentation are required:

CAPSULE ANSWER

A QS-9000 system is a documented system. Quality methods are planned and written down; results are recorded so that conformity can be objectively verified.

1. Policy documents that express management's methods for achieving and maintaining quality.
2. Design-related documents.
3. Instructional documents that tell people how to perform quality-related tasks.
4. External documents, from customers and others, that affect quality policies and practices.

Records must verify that requirements have been met (Question 33), and the documents described below must be retained and given to auditors upon request.

POLICY DOCUMENTS

- A quality policy statement (Question 22), defined and documented by management, that spells out the company's quality policy and goals. Management must make sure that the policy is understood by everyone in the company.
- A *Quality Manual* (Question 35), which is a brief, formal document that:
  - Describes the structure of the documented quality system.
  - Explains the company's quality policies as they relate to the QS-9000 standard.

DESIGN-RELATED DOCUMENTS

- Drawings, blueprints, and similar design-stage proofs (including CAD files, tooling drawings, gage prints, and so on).

- Quality planning documents, such as Failure Mode and Effects Analyses (FMEAs) (Question QUALPLAN).
- Production Part Approval Process (PPAP) documents (Question 39), including PPAP submissions and warrants.

INSTRUCTIONAL DOCUMENTS

- Procedures (sometimes called Quality Procedures, Standard Operating Procedures, Quality Practices, or Level 2 documents; see Question 36). These relatively general and functional documents describe, for quality-related processes:
  - Responsibility and authority.
  - General process steps.
  - Related documents.

  Procedures contain more detail than the *Quality Manual* and less detail than work instructions.
- Work instructions (sometimes called Level 3 documents). These task-level, step-by-step, "how-to" documents tend to be very detailed and specific. They are indeed "working documents," being very often in active and regular use at employee workstations. (QS-9000 requires that such documents be situated at work centers.) Level 3 documents include:
  - Operator instructions.
  - Control plans.
  - Inspection plans.
  - Setup instructions.
  - Bills of material.

EXTERNAL DOCUMENTS

- Customer specifications.
- Externally published standards (from ASTM, ISO, and similar bodies) that affect quality or the quality system.

A WORD OF CAUTION

Experience has shown that companies striving for QS-9000 registration do not have trouble assembling enough documents to meet the requirements of the Standard. Rather, they tend to overdocument the system—to pile on the paper.

QS-9000—and especially its parent, ISO 9000—vigorously oppose such non-value-added activity. These guidelines for determining the need for documents are given:

> . . . the range and detail of the procedures that form part of the quality system depend upon the complexity of the work, the methods used, and the skills and training needed by personnel involved in carrying out the activity. (4.2.2)

> . . . documented procedures [define] the manner of production, installation, and servicing, where the absence of such procedures could adversely affect quality. . . . (4.9.a)

These are important commonsense provisions. They allow some latitude as to the detail and quantity of the procedures a company develops. The pyramidal document hierarchy recommended in ISO 9000 is represented below. The *Quality Manual* is the most important component. If, for example, a person performing a job has a high level of qualifications, training, and experience in carrying out his or her task, the procedures governing that task need not be as detailed as they would be for someone who is less skilled.

APPLYING THE STANDARD

When making a judgment call like this one, be prepared to justify it to the QS-9000 auditor.

Often, companies implementing QS-9000 adapt their existing documents to meet the requirements of the Standard. When doing this, a zero-based

ISO 9000 Document Hierarchy

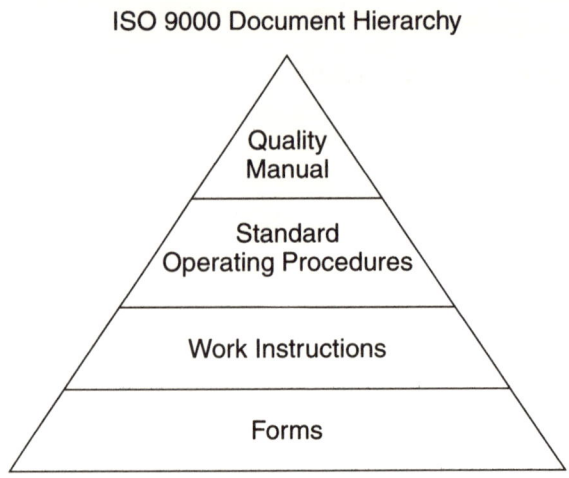

approach should be used. To make a document prove that it has to be in a quality system, use the "smell test" in Appendix D.

Printing the documents is not enough. A company is also required to "effectively implement documented procedures and the quality system" (Question 74). The auditor expects to see procedures in active use in the facility, not just stacked in a binder.

TECHNICAL REQUIREMENTS

1. A company must prepare a *Quality Manual* that:
   - Addresses the requirements of the Standard.
   - Outlines the structure of the documented system.
   - References quality system procedures.
2. Documented procedures covering tasks that affect quality must effectively meet:
   - The requirements of the Standard.
   - The company's stated quality policy.
3. The degree of required documentation depends on:
   - The methods used.
   - The skills needed.
   - The training acquired by the personnel who do the job.
4. Document control must be exerted over all quality system documentation (Question 38).

TECHNICAL GUIDELINES

A company should document—in a systematic and orderly manner, in the form of written policies and procedures—all elements, requirements, and provisions adopted for its quality system. This documentation should ensure common understanding of quality policies and procedures.

PURPOSE OF DOCUMENTATION

Process documentation is objective evidence in an audit that:

- A process has been defined.
- Procedures are approved.
- Procedures are under change control.

Preparation and use of documentation are meant to be "dynamic, high value-adding activities." They are essential for:

APPLYING THE STANDARD

Many European quality system experts feel that American firms overdocument their operations.

- Achieving required quality.
- Evaluating the quality system.
- Maintaining improvements.
- Making quality improvements.

DOCUMENTATION SYSTEM

All elements, requirements, and provisions in the quality system should be defined and documented. This documentation includes:

- *Quality Manual*, comprised of:
  - Quality policy.
  - Quality objectives.
  - Organizational structure.
  - Description of quality system.
  - Quality practices.
  - Structure and distribution of quality documentation.
- Quality plans, describing specific quality practices, resources, and sequence of activities.
- Procedures—written statements that specify:
  - The purpose and scope of activities.
  - How they are conducted, controlled, and recorded.

The documentation system must also include work instructions, specifications, and drawings.

OPERATIONAL PROCEDURES

These procedures, often called standard operating procedures (SOPs):

- Should be developed, issued, and maintained to:
  - Implement a company's quality policy and quality objectives.
  - Coordinate the different activities included in an effective quality system.

- Should state simply, unambiguously, and understandably:
  - The objectives of various activities that impact on quality.
  - The methods to be used.
  - The criteria that must be satisfied.
- Should represent separate work phases.
- Should include documentation of interfaces between work phases.

WORK INSTRUCTIONS

This element of the documentation system should be used to specify production operations to the appropriate extent, and to document common practices that occur throughout the facility.

Work instructions should:

- Describe criteria for determining:
  - Satisfactory work completion.
  - Conformity to specification.
  - Standards of good workmanship, defined to appropriate extent by written standards, photographs, and/or physical samples.
- Be available and/or accessible to the workers who need them.

REQUIRED DOCUMENTS

See above.

AUDIT ISSUES

The first step in a registration audit is usually a "desktop study" (review) of the *Quality Manual* and, often, the procedures. This step verifies that the documented system meets the requirements of the Standard. Thereafter, assessors check on the extent to which the documented system is understood and followed by affected employees.

Conformity of these documents to QS-9000 requirements is not usually a huge problem, but effective control of them can be (Question 38).

35. What is the purpose of a *Quality Manual?*

The *Quality Manual*, required by the Standard, is a relatively short, controlled-circulation document that describes:

> **CAPSULE ANSWER**
>
> The *Quality Manual* is the principal document that describes the facility's quality system, quality policy, quality commitment, and documentation structure.

■ A company's quality policy.
■ The policies it follows to meet the requirements of QS-9000.
■ The structure of its quality system.

How short is "short"? Forty pages is typical; fifty can be too many. In the world of QS-9000, where companies routinely create manuals that are indeed "doorstops and footstools," this may seem like a radical notion. How can a good *Quality Manual* be kept this short? By describing only policies.

The manual does not tell how to do anything; it only talks about what the policies are. Verbiage is restricted to the requirements of the Standard and an occasional related vital point. Otherwise, the "how to" is reserved for the procedures (Question 36) and work instructions (Question 37).

The *Quality Manual* serves several important purposes:

■ Management uses it to state its quality policy (Question 22) and affirm its commitment to that policy.
■ A quality system audit path is created.
■ As a marketing tool, it documents the company's quality policy to potential customers. For this reason, no confidential or proprietary material should appear in the manual.
■ It acts as a training tool for new employees.
■ During management reviews, it functions as the quality system "bible."

TECHNICAL REQUIREMENTS

1. The *Quality Manual* must address the requirements of the Standard.
2. The *Quality Manual* must describe the structure of the quality system documents, and reference relevant procedures.

TECHNICAL GUIDELINES

- Typically, the *Quality Manual* is a stand-alone document. This makes it easy to circulate and update as required. In very small organizations, the *Quality Manual* may include the quality system procedures.
- The *Quality Manual* can take various forms, depending on the size of the company, the nature of the activity, and the scope of operations.
- Larger organizations may have a "corporate" quality manual, supported by divisional quality manuals and specialized quality manuals (as needed).
- Because the Standard does not prescribe the structure of a *Quality Manual*, each company has great flexibility.
- The content of the manual should stick close to the requirements, but it is not necessary to use the same sequence or structure. Nor is it necessary to mimic the Standard. The manual should be expressed in the company's own language and style.
- For guidance, users can consult ISO 10013, *Guidelines for Developing Quality Manuals*. Not that this is a guidance document only.

REQUIRED DOCUMENTS

1. *Quality Manual.*

AUDIT ISSUES

The first step in a registration audit is usually a "desktop study" (review) of the current *Quality Manual*. This is done to verify that the documented system meets the requirements of the Standard. Because the manual is the first thing that assessors see about a company, every effort should be made to make it perfect.

36. What kinds of procedures are required by the Standard?

Procedures are the main instructional and operational documents of the

CAPSULE ANSWER

Procedures spell out how quality-related processes are carried out.

QS-9000 quality system. They define the approved, "best" ways to carry out quality-related processes.

Of the 23 main elements of QS-9000, 16 specifically require procedures. Four require procedures under certain conditions. The others do not specifically call out procedures, but it is advisable to have them (Appendix E).

A documented system will have, at a minimum, 23 procedures. Some companies have around 26. Having more than 26 is usually not desirable.

Procedures must be understood and followed by everyone they apply to. Procedures must be living documents that are revised and updated to keep them consistent with best practices.

> **APPLYING THE STANDARD**
>
> A *process* is a series of tasks carried out by a number of functions. A *task* is a series of actions performed by one function.

WHAT IS A PROCEDURE?

A procedure (sometimes called a standard operating procedure, a quality procedure, or a Level 2 document) defines the steps needed to carry out quality-related processes. Procedures are directly linked to relevant sections of the *Quality Manual* (Question 35), which in turn are directly linked to relevant clauses of QS-9000.

Procedures must not be confused with work instructions (Question 37), which are Level 3 documents that define how specific, single-function tasks are carried out. Procedures define processes. For example, changing a tire is a process, which is made up of a series of tasks. We might write a procedure for how to change a tire. Then we might write work instructions for specific tasks within the process (e.g., assembling the jack).

HOW SHOULD A PROCEDURE BE STRUCTURED?

A great deal of flexibility is allowed in structuring a procedure (see page 91). When a suitable structure is found, a company should follow it consistently. At a minimum, a procedure must include:

■ The titles of the function(s) responsible.
■ The name of the relevant *Quality Manual* clause.

Comparison of Documentation Levels: Inspection and Testing

Quality Manual Section 10 (Ref.: ISO 9001 4.10.2)	Standard Operating Procedure 10 (Reference: Quality Manual)	Receiving Inspection Work Instruction (Reference: SOP-10)
No incoming product is used until it has been inspected in accordance with procedures and found to conform to specified requirements. Where incoming product is released for urgent purposes prior to inspection, it is positively identified and recorded.	The Warehouse Manager is responsible for carrying out receiving inspection. He inspects incoming product against the specifications on the Purchase Order and Purchase Traveler, and records his findings on the latter. When incoming product passes inspection, a copy of the signed-off Purchase Traveler is affixed to the product, and another copy is routed to Accounting. Where incoming product is found to be suspect or nonconforming, it is diverted into the Quarantine Area, where it is processed in accordance with SOP-13. Where incoming product must be released into production prior to inspection, the Warehouse Manager affixes a red Traceability Tag to each component. When such product is found to be defective or otherwise nonconforming, production personnel report same to the Warehouse Manager. He immediately recalls the lot.	Responsibility: Warehouse Clerk. 1. Pull delivery copy of shipper. 2. Pull purchase order copy from Incoming Shipments bin. 3. Fill out top section of Purchase Traveler with relevant information. 4. Compare characteristics on Purchase Order with shipment. Note results on Purchase Traveler. Check off "conforming" or "nonconforming" in appropriate box. 5. When nonconforming, route lot to Quarantine Area. Submit Purchase Order, Purchase Traveler, and Shipper to Quality Department. 6. When inspection is complete and material is found to be conforming, affix copy of Purchase Traveler to lot. Notify intended department of receipt. 7. Send other copy of Purchase Traveler, Shipper, and Purchase Order to Accounting.

- The steps needed to complete the process or processes covered.
- Certain mandatory document control features (Question 38).

It is also advisable to include:

- A list of relevant work instructions.
- A list of controlled records.
- A list of other documents needed to complete the procedure.
- A brief instruction on how to carry out corrective action if problems arise in following the procedure.

How Should a Procedure Be Developed?

The best way is to start with an existing document. In most companies working toward QS-9000, there is already a sizable (sometimes too sizable) body of procedures and similar documents. Find one that is relevant and adapt it. Failing that, create a document "from scratch."

With either approach, it's best to have the "doers"—the people who are subject to the procedure and are most familiar with the process(es) involved—assigned to the initial writing, as well as any edits and updates that are needed before the final product is approved for issue.

A procedure must document what a work unit actually does. And what it does must be consistent with the requirements of the Standard. Any review of a procedure during the development stage must take into account the requirements of the Standard.

What Happens after a Procedure Is Written?

It is implemented (Question 74). The people who are subject to the procedure must be educated as to:

- Where controlled copies of the procedure are located.
- What the procedure says.
- How to request changes to the procedure whenever changes are needed.

Implementation is confirmed, first, by internal quality audits, and then by external audits carried out by registration assessors.

It is important to keep in mind that these are "living documents." They are never "done." They must change as a business changes, as the needs of customers change, and as new practices are developed and perfected.

TECHNICAL REQUIREMENTS

1. A company must prepare procedures that are consistent with:
   - The requirements of the Standard.
   - The company's own quality policy.
2. The company must effectively implement the procedures (Question 74).

TECHNICAL GUIDELINES

The scope of the procedures, and the amount of specifics in them, should be consistent with:

- The complexity of the work.
- The methods used.
- The skills/training needed by the functions involved.

Procedures should reference work instructions (usually, written) to define how specific tasks are carried out (Question 37).

REQUIRED DOCUMENTS

1. Procedures, as described above.

AUDIT ISSUES

The *Quality Manual* is a key part of the desktop study at registration audit time. The on-site examination and internal audits focus on how effectively the procedures are being implemented.

Both types of auditors assess:

- How well employees understand the procedures that apply to them.
- Whether employees understand how to effect changes to procedures.
- How well procedures are controlled (Question 38).

37. What kinds of work instructions are required by the Standard?

A work instruction is sometimes called an "operator instruction," "standard operating sheet," or "Level 3 document." One form of work instruction is what QS-9000 calls "process monitoring and operator instructions." QS-9000 requires these; ISO 9000 requires work instructions only under certain circumstances.

Typically, work instructions describe in detail how production, monitoring, and inspection tasks are carried out. The instructions, which can take many different forms, depending on how they are used, must be available to employees at their workstations. Work instructions may also define the steps for carrying out other, non-production-related tasks.

Work instructions need not be text-based. They can be flow charts, pictures, or diagrams. Because work instructions are directed toward production employees, illustrations or diagrams can be much more effective and are highly recommended.

WHAT IS A WORK INSTRUCTION?

A work instruction defines the steps needed to complete a quality-related task, which is almost always carried out by a specific function or employee. Work instructions are derived from relevant standard operating procedures (SOPs) (Question 36). These originate from relevant sections of the *Quality Manual* (Question 35), which in turn is directly linked to the relevant clause of the QS-9000 standard.

Work instructions must not be confused with procedures (Question 36), which are Level 2 documents that define processes. Work instructions spell out how specific, single-function tasks are carried out.

HOW SHOULD A WORK INSTRUCTION BE STRUCTURED?

Work instructions can look very different within a single company, because they accomplish many different types of tasks. This diversity requires a great deal of flexibility in the formats used.

Typical work instructions include:

- Dimensional control plans.
- Gage instructions.
- Inspection control plans.
- Inspection instructions.
- Laboratory test instructions.
- Operator instructions.
- Part number cross-references.
- Process sheets.
- Traveler sheets.

For each type of work instruction used, a company should set a standard format and follow it consistently.

At a minimum, the procedure must include:

- The titles of the function(s) responsible.
- The name of the relevant procedure.
- The steps needed to complete the task.
- Certain document control features (Question 38).

It is also advisable to include:

- A list of related work instructions.
- A list of controlled records.
- A list of forms and other documents needed to complete the task.
- A brief instruction on corrective action, if problems arise in following the procedure.

How Should a Work Instruction Be Developed?

The best approach is to start with an existing document. Many companies in the Big 3 supplier community already have extensive work instructions and related documents. Find one that is relevant and adapt it. The alternative is to create the document "from scratch."

Whichever approach is used, it's best to have the "doers"—i.e., the people who are subject to the procedure and are most familiar with the process(es) involved—do the initial writing and any necessary edits and up-dates, before the final product is approved for issue.

A work instruction must document the tasks actually performed. And those tasks must be consistent with the requirements of the related proce-dure and QS-9000. Therefore, any review of a work instruction during the development stage must take into account what is required by the proce-dure, the *Quality Manual*, and the Standard.

WHAT HAPPENS AFTER A WORK INSTRUCTION IS WRITTEN?

It is implemented (Question 74). A successful implementation requires ed-ucating the people who are subject to the instruction as to:

- Where controlled copies of the work instruction are located (at work sites, as required).
- What the work instruction says.
- How to request changes to the work instruction whenever changes are needed.

Implementation is confirmed by internal quality audits and, later, by exter-nal audits carried out by registration assessors. As a practical matter, how-ever, registration assessors rarely go into detail on the content of work instructions.

It is important to consider these as "living documents" that are never "done." They must change as a business changes, as the needs of customers change, and as new products, processes, and practices are developed and perfected.

TECHNICAL REQUIREMENTS

1. The company must prepare *written* process monitoring and operator instructions for all employees who operate processes.
2. Instructions must be available at workstations.
3. Instructions must include or make reference to the following, as appropriate:
   - Corrective action instructions.
   - Current engineering date.
   - Current engineering level.

- Engineering and manufacturing standards, as relevant.
- Inspection instructions.
- Instructions for identifying and disposing of material.
- Operation name.
- Operation number (linked to process flowchart).
- Part name.
- Part number.
- Required gages, tools, and equipment.
- Revision approvals.
- Revision date.
- Statistical process control (SPC) requirements.
- Special characteristics designed by the customer and the producing company (Question 49).
- Tool change intervals.
- Tool setup instructions.
- Visual aids.

TECHNICAL GUIDELINES

Work instructions should draw on the sources listed in the *Advanced Product Quality Planning and Control Plan* manual (Question 30).

REQUIRED DOCUMENTS

1. Work instructions, as described above.

AUDIT ISSUES

In many companies, the number and profusion of work instructions make it impossible for registration auditors to audit them comprehensively. They do, however, check for their existence, scan some of them for appropriateness, verify the effectiveness of document control, and satisfy themselves that employees are aware of the work instructions and follow them.

Internal and external auditors assess:

- How well employees understand the work instructions that apply to them and the procedures for introducing changes.
- How well the work instructions are controlled (Question 38).

38. What are the requirements for document control?

All documents used to control quality and the process must be controlled. The principle of control is deceptively clear: The system must ensure that the latest approved versions of key documents are reasonably available to those who need them. The reality is considerably dimmer: This element raises more noncompliances than any other element of the Standard.

These are the types of documents that must be controlled:

1. Internally created documents:
   - Business plan (Question 24).
   - *Quality Manual* (Question 35).
   - Standard operating procedures (Question 36).
   - Work instructions (Question 37).
   - Standard forms.
   - Product drawings, blueprints, computer-aided design (CAD) files, and similar graphics.
2. Externally generated documents that affect quality:
   - Customer engineering standards and specifications.
   - National and international quality, manufacturing, and design standards (ISO, ASTM, NIST, and so on).
   - Customer-mandated drawings, guidebooks, manuals, and the like (AIAG, and so on).

CONTROLLING INTERNALLY CREATED DOCUMENTS

A company is required to have a documented procedure for how these documents are:

- Developed and approved.
- Distributed in a controlled fashion to ensure that they are reasonably available to the people who need them.

- Amended on a planned, disciplined basis, with appropriate reviews and approvals.
- Withdrawn in a systematic manner when rendered obsolete.

A typical control system includes, for each document:

- A master copy, with original approval signatures.
- Controlled copies, made from the master and stamped or marked "CONTROLLED," indicating the current revision number and/or date.
- A controlled circulation list that itemizes the functions and locations where controlled copies are placed.
- A circulation sign-off, proving that updated documents were issued.
- An amendment history, briefly summarizing the changes that have been made over time.
- An archive of obsolete issues, clearly marked "OBSOLETE," for reference purposes.

The procedure should describe the method for changing documents as needed. *All employees* should be empowered to request or suggest changes and improvements to quality system documents. This is not a requirement of the Standard, but common sense and implementation experience make it a must.

CONTROLLING EXTERNALLY GENERATED DOCUMENTS

The procedure here must provide for:

- Designation of a "master copy."
- A method of verifying, on a defined basis, that the master copy is the most recent issue. (With externally published standards and specifications, this can involve subscribing to an automatic update service, or an annual query to publishers for verification that the edition on hand is the most recent one.)
- A controlled circulation list with signoffs, proving that updated documents were issued.
- As an alternative to the above list, log-in/log-out sheets and records, to track the whereabouts of master and controlled copies.

■ An archive of obsolete issues, clearly marked "OBSOLETE," for reference purposes.

WHAT ABOUT UNCONTROLLED CIRCULATION?

Uncontrolled copies of controlled documents may be issued for specific, time-limited purposes. Uncontrolled copies escape from the document control system; they are not retrieved or updated. Therefore, where quality-related activities are taking place, uncontrolled copies must never be issued or used in place of controlled copies.

But, uncontrolled circulation has its uses. For example, at a training session on a new procedure, uncontrolled copies can be distributed and then collected and destroyed when the class is finished. Or, when a customer asks for a copy of a *Quality Manual*, an uncontrolled copy can be given to satisfy the request.

WHAT ABOUT COMPUTER-BASED DOCUMENT SYSTEMS?

Utilizing PC networks, mainframes, and/or CAD networks to control certain quality system documents can save time and headaches. However, such systems are not panaceas. They are most practical for controlling:

■ Design-related documents (most of these are CAD-based anyway).
■ The *Quality Manual*.
■ The more general procedures.

As a practical matter, specific procedures, work instructions, and other quality system documents tend not to be amenable to computerization because many organizations do not have networks that are comprehensive enough to permit adequate access to all employees.

Computerized document control systems are best implemented when they already exist for other purposes. A common pitfall in QS-9000 (and ISO 9000, for that matter) is to try to implement QS-9000 AND a computerized document control system at the same time. The double-lane approach is just too burdensome and frustrating.

TECHNICAL REQUIREMENTS

The required document control system must include the following elements:

1. Control of all documents and data (hard copy or electronic) that pertain to QS-9000 requirements. They include, as applicable, documents of external origin, such as published standards; customer drawings; engineering specifications; and documents referenced by customer drawings or specifications.

> **APPLYING THE STANDARD**
>
> Copies of QS-9000 must be controlled!

2. Utilization of the customer's special characteristic symbols (Question 49) on all process-related documents.
3. Review, and approval for adequacy, of all quality system documents by authorized personnel prior to issue.
4. A master list (or an equivalent document-control procedure) identifying the current revision status of documents, to preclude the use of invalid and/or obsolete documents.
5. Review, and approval for adequacy, of all quality system document revisions, by the same functions that are responsible for the original review and approval, unless specifically designated otherwise.
6. Review, distribution, and implementation of all customer engineering drawings, standards, specifications, and changes. "Timely" scheduling should be expressed in business days, not weeks or months.

The quality system must include controls to ensure:

1. Availability of relevant editions (that is, the correct current editions) of quality system documents wherever operations essential to the effective functioning of the quality system are performed.
2. Prompt removal of invalid and/or obsolete documents. If removal is impractical (for example, with certain computer files), documents must be protected from unintended use.
3. Suitable identification of any obsolete documents retained for legal and/or knowledge-preservation purposes.

TECHNICAL GUIDELINES

These features are recommended for a document control system:

- Pertinent current editions of documents should be available at places where work vital to the effective functioning of the quality system is performed.
- Procedures must set forth:
  - How documents are controlled.
  - Who is responsible for control.
  - What is controlled.
  - Where and when control takes place.
- Documents can be published in a variety of media.
- The mechanism for changing and updating documents should:
  - Ensure accurate updating.
  - Ensure that only authorized documents are used.
  - Preclude confusion.
  - Take into account the effect of the change on other parts of the system.

Controlled documents are those pertinent to:

- Design.
- Purchasing.
- Performance of work.
- Quality standards.
- Inspection of materials.
- Planning and progress reviews.
- Interactions of supplier with purchaser.
- Design/development inputs and outputs.
- Design plans, verifications, and results.
- Maintenance.

Documents that are in an effective document control system will be:

- Carrying an authorization status.
- Clean.
- Clear.
- Dated (including revision dates).
- Easily identifiable.
- Legible.

REQUIRED DOCUMENTS

1. A procedure addressing the document control system and its requirements.
2. Log sheets, as needed.

AUDIT ISSUES

Potential pitfalls here are many. The most common problems include:

- The infamous "sticky-note syndrome"—controlled copies marked up with unapproved changes, revisions, and marginal comments.
- Unapproved instructions in use; for example, gauging instructions taped to gages and fixtures.
- Obsolete documents kept or posted where they might be mistaken for current documents.
- Unmarked (and often unneeded) "archives" of obsolete documents.
- Documents that do not reflect actual practices because they have not been either effectively implemented (Question 74) or updated to reflect actual practices.

Relations with Customers and Vendors

39. What is the Production Part Approval Process (PPAP)?

This is a system for obtaining customers' approval of the production process for a particular part, product, material, or service. Production part approval must be obtained for:

CAPSULE ANSWER

Production part approval is required for all parts, services, and materials, except as waived by customers.

- Newly designed parts, services, or materials.
- Changes to previously approved submissions (under varying circumstances).
- Fulfillment of special directions from a customer.

It is strictly up to the customer to determine whether production part approval must be obtained for the item(s) a supplier is providing. However, to meet QS-9000 requirements, the supplier must have a procedure covering PPAP and must obtain written waivers from the customer if PPAP is not required.

The process is defined in a manual published by the Automotive Industry Action Group (AIAG): *Production Part Approval Process.*

Companies that have been supplying Tier 1 automotive facilities are familiar with PPAP. For them, meeting QS-9000 requirements is mainly a matter of documenting how they do PPAP in a procedure.

However, QS-9000 is bringing into the picture many companies that have not been exposed to PPAP. It is mandatory that these companies:

- Consult with their customer(s) to determine whether PPAP is required.
- Obtain a copy of the PPAP manual from AIAG (Question 18).
- Write a PPAP procedure for their QS-9000 system, to define the process in case a current or future customer requires PPAP.

Given below is a partial summary of the requirements drawn from the PPAP manual. This information is for reference only. *Any company that is developing a system and procedure should obtain the current edition of the manual for reference.*

TECHNICAL REQUIREMENTS

1. A company must comply with the requirements of the Production Part Approval Process (PPAP) manual.
2. Production part approval covers all production and service commodities, including bulk materials.
3. Production part approval is always required for each part, prior to the first quantity shipment, whenever:
    - A new part is introduced.
    - A discrepancy on a previous submission is found.
    - A product is modified by an engineering change.

4. A company must notify the customer, to determine whether production part approval will be required, whenever:
   - A material change from the previous approval is made.
   - A change, refurbishment, transfer, or rearrangement is made in (nonperishable) tools, dies, molds, or patterns.
   - The process or manufacturing method changes.
   - The source for subcontracted parts, materials, or services changes.
   - A product is rereleased after tooling has been inactive for twelve months or more.
   - A customer has requested that shipment be suspended due to a supplier's concern about quality.

5. The content of submission depends on the submission level assigned by the customer to a company (or, possibly, to the company and a part number). Submission levels are:
   - Level 1—Warrant only, submitted to customer.
   - Level 2—Warrant with product samples and limited supporting data.
   - Level 3 (default)—Warrant with product samples and complete supporting data.
   - Level 4—Warrant (no product samples) with complete supporting data.
   - Level 5—Warrant with product samples and complete supporting data to be reviewed by the customer at the company's manufacturing location.

6. PPAP submissions must include (as required by the customer and based on the customer-assigned submission level—see the chart on page 106):
   - Submission warrant (see PPAP manual).
   - Two sample parts.
   - Design records (where applicable).
   - Authorized engineering change documents, if any.
   - Dimensional results.
   - Checking aids.
   - Test results as specified in design records.
   - Process flow charts.
   - Process Failure Mode and Effects Analyses (PFMEA).
   - Design FMEA, if company is design-responsible.
   - Control plans.
   - Process performance evaluations.
   - Measurement system variation (gage R & R) studies for all equipment referenced in the control plans.

Retention/Submission Requirements	Levels				
	1	2	3	4	5
1. Warrant	S	S	S	S	R
2. Appearance approval report	S	S	S	S	R
3. Sample product/Master sample	R	S	S	R	R
	R	R	R	R	R
4. Design records	R	S	S	S	R
	R	S*	S*	S*	R
5. Change documents (if any)	R	S	S	S	R
6. Dimensional results	R	S	S	S	R
7. Checking aids	R	R†	R†	R†	R
8. Test results	R	S	S	S	R
9. Process flow charts	R	R	S	S	R
10. Process FMEAs	R	R	S	S	R
11. Control plans	R	R	S	S	R
12. Process performance	R	R	S	S	R
13. Measurement system studies	R	R	S	S	R
14. Design engineering approval	R	R	S	S	R

S = Submit to customer.
R = No submission required; retain on site.
*Unless waived by customer.
†Submit when required by customer.

7. A company must notify the customer (and, possibly, resubmit a production part for approval), whenever there is a change in:
   ■ Engineering level.
   ■ Manufacturing location.
   ■ Material subcontractor.
   ■ Part number.
   ■ Production process environment.
8. A company must validate such changes in accordance with PPAP.
9. A company must document its PPAP process in a procedure. (Note: This is not in the Standard, but is an interpretation rendered in a QS-9000 assessment.)

REQUIRED DOCUMENTS

1. PPAP Procedure.
2. PPAP records and/or waivers.

40. Are sales and order-taking activities covered by the Standard?

CAPSULE ANSWER

The process of creating and accepting an order between customer and supplier must be controlled and documented.

Yes. The Standard refers to the sales/order-taking process as "contract review." This does not necessarily imply or require a contract that is printed, signed, and witnessed, nor does it require doing business face-to-face with a customer. Verbal orders are not precluded; the Standard defines "contract" and "accepted order" as "agreed requirements between a supplier and a customer transmitted by any means." The Standard uses the term "contract" to denote the nature of the agreement between the supplier and the customer.

In some companies, contract review is as simple as looking at a customer order, verifying the information, signing off, and shooting it through. In others, especially companies that get involved in intricate and long-term bidding, contract review may be an extended process involving many functions and steps.

For companies that are design-responsible, contract review may be closely linked with Advance Quality Planning (Question 30), Design Control, and other elements of the quality system. These links must be clearly delineated.

Contract review is the first (and, arguably, the most critical) opportunity for preventive action—the first and best chance to "get it right the first time." It is the stage at which the supplier–customer agreement is defined and clarified. The more thorough and detailed the definition, the better. In essence, the Standard requires understanding and agreement on:

- What is to be provided.
- The terms of the transaction.
- Systems for making changes.
- Systems for settling disputes.

The Standard also requires strict control and thorough documentation of the contract review process.

TECHNICAL REQUIREMENTS

1. A documented procedure covering order-taking activities is necessary.
2. The company must "review each statement of requirements" (which means every order, contract, purchase order, and sales agreement) to ensure that:
   - The customer's requirements are adequately defined and documented.
   - The company has the capability to meet the customer's requirements.
3. Should anything about the order be inaccurate, unclear, or outside company capability, the company must:
   - Resolve the matter with the customer.
   - Communicate any order changes to the appropriate functions within the organization.
4. Verbal orders are acceptable, as long as the company and the customer agree to the requirements before the order is accepted.
5. Order changes must be effectively communicated to affected functions.
6. The company must maintain records of contract reviews.

> **APPLYING THE STANDARD**
>
> Verbal orders should make reference to specific written terms, and the supplier must be able to prove that the terms were reviewed and agreed to before order acceptance.

TECHNICAL GUIDELINES

A company should establish clearly stated and understood channels for communication and "interface" with a customer.

The following elements should be included in contract reviews, as applicable:

- Opportunities for all parties to review requirements and resolve questions.
- Assignment of responsible individuals on both sides.
- Acceptance criteria.
- Definitions of critical terms.
- Identification of risks.

- Clear descriptions of functional requirements, including all aspects needed to satisfy the customer's needs for performance, safety, reliability, and security.
- Protection of proprietary information.
- Resolution of any customer requirements that differ from the supplier's capabilities.
- Definition of the supplier's responsibilities with respect to subcontracted work.
- Inclusion of draft quality plan, if appropriate.
- Process for addressing changes in customer requirements.
- Review and acceptance of test results.
- Customer obligations, including (as applicable) product installation and acceptance, as well as facilities and tools required of customer.
- Standards and procedures to be used.
- Review of requests for quotations, bids, and other preorder documents.

REQUIRED DOCUMENTS

1. Contract review procedure.
2. Contract review records.

AUDIT ISSUES

Sales and order-taking requirements are normally not difficult to meet, although companies with design responsibility may find them to be complex. Where internal or external salespeople have contract review involvement or responsibility, care must be taken to ensure that all of them are following the prescribed procedure. (Tip: Involve them in developing the procedure from the outset.)

A common audit trap is failure to define how order changes are effectively communicated.

41. Besides sales and order taking, what other customer interactions does the Standard address?

The Standard requires care of customer-owned items and materials, and a formal, documented system for assessing customer satisfaction.

> ## CAPSULE ANSWER
>
> The Standard requires controls over customer-owned property provided for company use. It also requires a documented process for assessing and tracking customer satisfaction.

CUSTOMER-SUPPLIED PRODUCT

A customer may provide a supplier with products or materials that it owns, on which the supplier is to take some agreed-on action. For example, a customer will often provide a supplier with materials that the supplier is to incorporate into the final product that is sold to the customer. To illustrate:

- An assembly plant (the customer) sends fasteners to a component manufacturer (the supplier). The manufacturer uses the fasteners in the component it is furnishing to the assembly plant. The fasteners remain the property of the assembly plant; they are "customer-supplied product."
- A Tier 1 plant (the customer) provides returnable containers to its Tier 2 supplier for use. The containers are "customer-supplied product."

At times, a customer will provide tooling or test equipment to a supplier. This too is "customer-supplied product" and must be controlled.

The Standard requires a documented procedure for the control of customer-supplied material. The company must report to the customer any loss, damage, or nonconformity in such material.

CUSTOMER SATISFACTION

The Standard also requires that the company implement procedures for:

- Determining customer satisfaction.
- Tracking trends.
- Comparing and analyzing trends.
- Conducting reviews by senior management.

This requirement is a departure from ISO 9000. The company is compelled to focus its efforts (and its quality system) on customer needs: determining them, fulfilling them, and analyzing how well they are being accomplished.

TECHNICAL REQUIREMENTS

1. The Standard requires a documented process to control any product received from customers for any reason (as described above). The procedures must provide for:
   - Verification that the items are in fact those that the customers identified.
   - Storage under safe and controlled conditions.
   - Maintenance with the same standard of care given to the company's own property.
2. Any problems must be reported promptly to the customer.
3. A company must have a documented process for determining customer satisfaction. The process must include:
   - How satisfaction is determined.
   - The frequency with which satisfaction is determined.
   - A trend analysis supported by objective information.
   - A comparison of trends with other identified indicators or benchmarks.
   - A review by top management.

TECHNICAL GUIDELINES

The following measures are suggested for controlling customer-supplied product:

- Establish a documented reporting procedure with the customer.
- Examine the product, upon receipt, to:
  - Verify identity.
  - Verify quantity.
  - Inspect for damage.
- Inspect periodically for signs of deterioration.
- Protect the product from unauthorized use or disposal.

- Incorporate into the maintenance and use activities any limitations on storage time.
- Reinspect as required by contract.
- Adopt procedures for the protection of customer property acquired during the delivery process.
- Ensure that any equipment provided to the customer is suitable for the purpose(s) and is accompanied by written instructions for use, as appropriate.
- Establish a customer satisfaction system that takes into account immediate and ultimate customers, as appropriate.

REQUIRED DOCUMENTS

1. Procedure for control of customer-supplied product, with supporting records (where relevant).
2. Procedure for customer satisfaction, with supporting records (where relevant).

42. What are the requirements for after-sale servicing?

Servicing is a rather vague and indeterminate subject area of QS-9000. Its provisions apply when "servicing" is part of the original sales contract—that is, when service is sold along with a product.

Service may be in the form of an after-sales warranty or an actual service

> **CAPSULE ANSWER**
>
> Provide installation and servicing assistance as agreed by contract with the customer. Meet all specified requirements in accordance with documented procedures that are supported by records.

contract. The latter could require a company to provide on-site service people. Circumstances can vary. If a service requirement is not applicable to a company's activity, its *Quality Manual* should prominently display a statement to that effect.

To the extent that customers require service, the service practices must be documented in a procedure. Although no specific record keeping is called for in the Standard, an appropriate amount of documentation should be kept as proof of compliance.

TECHNICAL REQUIREMENTS

1. Where servicing is included as a requirement in a customer contract, establish and maintain documented procedures for:
   - Performing contracted services.
   - Verifying that the services meet the contract requirements.
   - Documenting the services as proof of performance.
2. The procedure must include methods for communicating servicing concerns to other process activities such as design, engineering, and manufacturing.

TECHNICAL GUIDELINES

Document installations of product or equipment with procedures and warning notices, where applicable. Include provisions that preclude improper installation or factors that could degrade:

- Quality.
- Reliability.
- Safety.
- Product performance.

Servicing is often a contract requirement in situations where functionality depends on regular maintenance. Specify:

- Who provides the service.
- What activities are provided.
- What tools and equipment are needed.
- What measuring and test equipment is needed.
- What documentation and instructions are needed.
- Requirements for backup, technical support, and spares.
- Qualifications of servicing personnel.
- Means by which customer feedback is channeled so as to influence design and redesign of products and services.

Other servicing activities are:

- Validate any necessary special-purpose tools or equipment.
- Control inspection, measuring, and test equipment used in the field.

- Document procedures and instructions for field activities, including verification and suitability of instructions for intended readers.
- Ensure adequate logistic backup regarding technical advice, materials, and equipment.
- Define all responsibilities clearly and in detail.
- Provide customers with end-use application information for correct implementation and handling.
- Provide proper advice by obtaining the most detailed information available.
- Control measuring and test equipment used in field installation and testing.
- Where necessary, provide instructions that are comprehensive, timely, and suitable for the intended readers.
- Provide assistance for adequate logistic backup, including technical advice and competent servicing.
- Assign and agree on a clear distribution of responsibilities among suppliers, distributors, and customers.

REQUIRED DOCUMENTS

1. Servicing procedure (where relevant).

AUDIT ISSUES

If a servicing requirement is not applicable to its activity, the company should be prepared to justify this to the assessors.

43. How do vendors figure in a QS-9000 quality system?

QS-9000 requires close control of vendor selection and the purchasing process itself (Question 44).

These requirements apply to the purchase of products and services that have an impact on the quality of a company's output. In that way, QS-9000

> **CAPSULE ANSWER**
>
> The Standard requires a documented process for selecting and controlling vendors—including "developing" their quality systems in conformity to QS-9000.

applies to suppliers of raw material. However, it also applies to suppliers of maintenance and repair products (and services—an element sometimes overlooked).

The Standard requires that vendors be chosen after they have been evaluated against defined criteria. Sometimes, a customer sends a list of suppliers to choose from (as with potential QS-9000 registrars; see Question 91). In that situation, pick vendors from the list. Otherwise, pick vendors only after checking them out.

After selecting vendors, a company must monitor their performance in the following ways:

■ Require 100 percent on-time delivery, and work with them to enable them to meet this requirement.
■ Help them develop their quality systems, using QS-9000 as the basic standard.

These measures will not be possible with all vendors, but the Standard expects companies to impress on their suppliers the need to work toward a common quality system standard (QS-9000) and to show evidence that this effort is underway. Suggestions include:

■ Audit vendors to QS-9000 requirements.
■ Encourage, exhort, or (where possible) require vendors to register to QS-9000 themselves.
■ Provide vendors with training in QS-9000 systems.
■ Where feasible, make QS-9000 adherence a requirement for vendor approval.

TECHNICAL REQUIREMENTS

1. Implement a documented procedure covering vendor selection and control.
2. Establish requirements for subcontractors, including (but certainly not limited to):
   ■ Quality system requirements.
   ■ Quality assurance requirements.
3. Evaluate the ability of potential vendors to meet the necessary requirements, and choose only vendors who can meet them.

4. When customers provide approved vendor lists, purchase relevant materials from vendors on those lists. (Use of such vendors does not relieve a company of the obligation to ensure the quality of subcontracted products or services.)

5. Define a plan for controlling vendors, taking into account:
   ■ The type of product.
   ■ The impact of the subcontracted product on the quality of the final product.
   ■ Evidence of subcontractors' previously demonstrated capability and performance, including quality audit reports and/or quality records.

6. Develop subcontractors' quality systems, utilizing QS-9000. This may include assessing subcontractors against the QS-9000 requirements.

7. Require 100 percent on-time delivery from subcontractors. To facilitate this requirement:
   ■ Give them planning information and purchase commitments.
   ■ Monitor their performance.

8. Establish and maintain quality records of acceptable subcontractors.

TECHNICAL GUIDELINES

■ To confirm that vendors have the ability to furnish products or services that meet QS-9000 requirements:
   – Conduct on-site assessments/evaluations of vendors' capability or quality systems.
   – Evaluate product samples.
   – Evaluate past history with similar suppliers.
   – Evaluate test results of similar suppliers.
   – Evaluate statistical data relevant to the consistency of the vendor's process.
   – Evaluate published experiences of other companies with the same vendor.

■ Assess vendor performance at intervals consistent with:
   – The complexity and technical requirements of supplied products.
   – Previous vendor performance.

■ Assessments should consider factors such as:
   – The status of the vendor's own quality systems.
   – Compliance of products with specific requirements.
   – Delivery arrangements.
   – Total cost.

- Assessment of methods can include:
  - On-site evaluation.
  - Evaluation of samples.
  - Past history with similar products.
  - Test results with similar products.
  - Published experiences of other companies with the same vendor.
- Establish close working relationships, including feedback systems, with vendors, for:
  - Continual quality improvement.
  - Quick settlement of disputes, and resolution of other routine and nonroutine matters.
  - Controls, procedures, and records for receiving inspections, as appropriate.
- Agree with vendors on quality assurance standards. The arrangements should meet the needs of the purchaser and should avoid unnecessary costs. Possible arrangements include:
  - Reliance on vendors' quality systems.
  - A requirement that vendors submit specific inspection/test data and/or process control records with shipments.
  - A promise by vendors to maintain 100 percent inspection/testing.
  - A promise by vendors to conduct lot acceptance inspection/testing by sampling.
  - Implementation by vendors of a quality assurance system such as QS-9000, to include periodic assessments of quality practices by the purchaser or by an objective third party.
  - Nonacceptance of vendor quality assurance and substitution of in-house receiving inspection and/or screening of nonconforming supplied product.
- Give the customer the chance to decide whether it wants to verify materials to be supplied by a vendor.
- When appropriate, conduct design or other reviews with the vendor; include any such requirement in the contract with the vendor.
- Encourage vendors to apply statistical methods to keep measurement processes in control.

REQUIRED DOCUMENTS

1. Procedure for vendor selection and control.
2. Records of vendor selection and approval.

AUDIT ISSUES

Make sure that all vendors whose products or services impact the quality of the final product are evaluated, approved, and placed on an approved supplier list. Don't forget commonly overlooked areas such as:

- Gage calibration services.
- Transportation suppliers.
- Temporary help agencies.
- QS-9000 consultants.

All functions that do purchasing should have access to the latest approved supplier list—and should use it!

44. What about the purchasing process?

QS-9000 has fairly strict rules for a vendor selection system (Question 43). Some of the rules are tough to live by, but the Standard's requirements for the purchasing process itself tend to be straightforward. There is nothing especially difficult here.

> **CAPSULE ANSWER**
>
> Purchasing must be carried out in accordance with a procedure that includes review/approval of documents before release, and precise identification of items ordered.

The requirements apply to the purchase of products *and services* that can affect the quality of the final output. A purchaser is required to:

- Document the purchasing process in a procedure.
- Make sure purchasing documents are sufficiently clear and complete.
- Review and approve purchasing documents before release.
- Obey governmental rules concerning hazardous, toxic, and restricted substances.
- Notify suppliers if their quality will be checked at their site.
- Give customers the right to check the quality of a supplier's product.

Nothing here is especially onerous, and most companies shooting for QS-9000 already have fairly well-defined purchasing systems. In some companies, however, there are really two purchasing systems:

1. The officially documented one, which usually comes fairly close to complying with the QS-9000 requirements.
2. The unofficial, informal one, in which anyone who wants anything can go out and buy it without following much of a system.

The latter scenario has caused major headaches for companies attempting to implement QS-9000.

TECHNICAL REQUIREMENTS

> **REMEMBER:**
>
> "Supplier, Facility" = The Company
>
> "Subcontractor" = The Vendor

1. Maintain a documented procedure covering purchasing activities.
2. Before releasing purchasing documents (e.g., purchase orders), review them for accuracy and completeness. Approve them only if they meet all specified requirements.
3. Purchasing documents must specify data clearly describing the product ordered, including (as applicable):
   - The type, class, grade, title, or other positive identification.
   - References to relevant editions of specifications, drawings, and similar documents.
   - The title, number, and edition of the quality system standard to be applied, if any.
4. In order to verify the conformity of a purchased product at a subcontractor's site, this intent must be specified in the purchasing documents. Required details include:
   - Verification arrangements.
   - Method of product release.
5. A customer may require, by contract, permission to verify the conformity to specified requirements of materials supplied for a product. This verification may occur at either the subcontractor's premises, before shipment, or the receiver's premises, before release into production. Such customer verifications do not:
   - Absolve either the receiver or the subcontractor from the obligation to deliver product that is acceptable to the customer.
   - Preclude subsequent rejection of the product by the customer.

6. A company's procedure must include a process ensuring compliance with governmental safety rules for handling of toxic and hazardous substances.

TECHNICAL GUIDELINES

A company's purchasing process should:

- Ensure that requirements for supplied products and services are clearly defined, communicated, and understood by vendors.
- Implement adequate and documented controls to ensure that purchased product conforms to specific purchasing and regulatory requirements, as applicable.
- Maintain quality records related to purchasing, to:
  - Ensure the availability of historical data for assessment of vendor performance.
  - Help detect quality trends.
- Maintain records of lot identification for traceability purposes, if appropriate.
- Include specific technical product requirements, including applicable editions of specifications, drawings, and so on.
- Include other product-specific information, as appropriate, such as:
  - Precise identification of grade.
  - Inspection instructions and applicable specifications.
  - Requirements for evidence of process control.
  - Packaging, labeling, transportation, and delivery timing requirements.
  - Lab method specifications and analysis instructions.
- Refer to other pertinent technical information, such as national or international standards and various tests, as appropriate.
- Include the revision status of referenced documents.
- Make sure purchasing documents are reviewed and approved by appropriate personnel.
- To prevent unintended use or installation of nonconforming received materials, use quarantine areas or other appropriate screening methods.
- Maintain receiving inspection characteristics based on:
  - Criticality of product.
  - Capability of vendor.

- Balance the costs of inspection against the consequences of inadequate inspection.
- Ensure that all necessary tools, gages, meters, and so on, are available and properly calibrated, and that personnel are adequately trained.
- Consider keeping samples from each lot for a set period of time.

REQUIRED DOCUMENTS

1. Purchasing procedure.
2. Purchasing forms.

AUDIT ISSUES

- Is everyone who does purchasing for the company following the approved procedure? Functions in the business may be purchasing items that affect the quality of the output, even though they are not following (and perhaps are not even aware of) an "official" purchasing system. Auditors look for these situations.
- Are all purchases being made from vendors on the approved supplier list?

Designing the Products

45. How is the design of products covered under QS-9000 and how does a company know it is design-responsible?

A company is design-responsible if it has the authority, for any product it ships to a Big 3 customer, to:

CAPSULE ANSWER

The design process must be planned and controlled by a documented procedure, and carried out by qualified functions whose interrelationships are defined.

1. Establish a new product specification.
2. Change an existing product specification.

(Companies that do not fit that definition can skip this section.)

The Standard requires that a design process must be a controlled process, covered by a documented procedure. This procedure must:

- Define the different functions that are involved in the design process, and how they relate to each other.
- Indicate how design-related information is communicated among the functions and reviewed as required.
- Describe how design activities (or projects) are planned and updated.
- Relate the ways design responsibilities are assigned.

Design personnel must be properly qualified (and documented in the same manner as all other functions that affect quality). They must be equipped with "adequate resources" and possess the roster of skills that QS-9000 considers "appropriate."

This section—Questions 45 through 50—discusses QS-9000's requirements and recommendations for design control, particularly:

- General design requirements.
- Guidelines for design development.

For requirements and guidelines on the following design-related issues, see the questions noted:

- Advanced quality planning (Question 30).
- Production Part Approval Process (PPAP) (Question 39).
- Design input and output (Question 46).
- Design review (Question 47).
- Design verification and validation (Question 48).
- Special characteristics (Question 49).
- Design changes (Question 50).

TECHNICAL REQUIREMENTS

1. Design activity must be covered by a documented procedure.
2. Management must prepare plans for each design and development activity, including definition of responsibility.
3. Design activities must be assigned to personnel who are qualified and are equipped with adequate resources.

4. Interfaces (organizational and technical) among different groups that affect the design process must be defined.

5. Necessary design information must be documented, disseminated, and regularly reviewed.

6. Appropriate resources for computer-aided design, engineering, and analysis must be utilized, or subcontracted under company direction. This arrangement can be waived by the customer, but objective evidence of the waiver is required.

TECHNICAL GUIDELINES

GOAL OF THE DESIGN FUNCTION

The goal of the design function is to translate customer needs into technical specifications for materials, products, and processes. These specifications should:

- Be unambiguous and clearly define characteristics important to quality, including:
  - Acceptance criteria.
  - Fitness for purpose.
  - Safeguards against misuse.
  - Dependability.
  - Serviceability.
  - Safe disposability.
- Specify measurement methods, testing processes, and acceptance criteria during design and production phases, including:
  - Performance target values.
  - Tolerances.
  - Attributes/features.
  - Testing and measurement methods, equipment, and computer software.
- Result in products or services that:
  - Are producible, verifiable, and controllable.
  - Give customer satisfaction at acceptable prices.
- Enable the organization to earn a satisfactory financial return.

MANAGEMENT RESPONSIBILITIES

Management should:

- ■ Define relationships and interfaces among:
  - – Research and development.
  - – Marketing.
  - – Purchasing.
  - – Quality management.
  - – Engineering.
  - – Materials technology.
  - – Production and manufacturing.
  - – Service groups.
  - – Facilities management.
  - – Warehousing and transportation.
  - – Communication groups.
  - – Information systems.
- ■ Assign design functions to personnel who:
  - – Are qualified.
  - – Have access to the information and resources necessary to complete the work.
  - – Ensure that all who contribute are aware of quality responsibilities.
- ■ Ensure that all concerned are aware of their responsibilities.
- ■ Establish time-phased design programs with checkpoints at which evaluation of products and processes will take place.
- ■ Organize a design planning process that:
  - – Defines design development management.
  - – Includes a development schedule.
  - – Addresses progress control.
  - – Defines organizational responsibilities.
  - – Manages interfaces among groups.
- ■ Ensure that the marketing function aids in:
  - – Determining and defining customer needs and expectations.
  - – Defining product requirements.
  - – Providing concepts, supported by data, that aid in producing products and services to defined specifications at optimal cost.
- ■ Design personnel should be qualified in the following skills, as appropriate:
  - – Computer-aided design (CAD).
  - – Computer-aided manufacturing (CAM).
  - – Design for assembly (DFA).

- Design for manufacturing (DFM).
- Design of experiments (DOE).
- Design failure mode and effects analyses (DFMEA).
- Finite element analysis (FEA).
- Geometric dimensioning and tolerancing (GD&T).
- Quality function deployment (QFD).
- Reliability engineering plans.
- Simulation techniques.
- Solid modeling.
- Value engineering (VE).

OBJECTIVES OF THE DESIGN PLANNING PROCESS

- Specify designs to a level of detail that permits verification (at the end of the process) that the design meets the specified requirements.
- Utilize design planning procedures that include:
  - Sequential and parallel work schedules.
  - Design verifications.
  - Evaluation of safety, performance, and dependability.
  - Plans for product measurement, testing, and acceptance criteria.
  - Assignment of responsibilities.
- Organize design work groups to be responsible for:
  - The information that should be received and transmitted.
  - Identification of sending and receiving groups.
  - Purpose of information.
  - Identification of transmission mechanism.
  - Records to be maintained.
- Formulate design plans to cover:
  - Project definition and statement of objectives.
  - Process or methodology for transforming purchase requirements or specifications into product.
- Plan the project organization to include:
  - Team structure, subcontractors, and material resources.
  - Development phases and project schedule.
  - Related plans.
- Define the design phases, including:
  - Inputs and outputs for each phase.
  - Verification process to be carried out.
  - Analysis of potential problems.

- Identify the methods employed to ensure that *all* activities are carried out correctly:
  - Rules and practices.
  - Tools and techniques.
  - Management configuration.
- Obtain input from marketing, such as:
  - Information on product or service problems in relation to customer experience and expectations.
  - Clues to possible design changes.
- Ensure that designs provide clear and definitive technical data addressing:
  - Procurement.
  - Execution of work.
  - Verification of conformance.
  - Safety, environmental, and other applicable regulations.
  - The company's own quality policy.
- Establish time-phased design and development programs with checkpoints appropriate to the nature of the product. Phases depend on:
  - Product application.
  - Design complexity.
  - Extent of innovation and technology.
  - Degree of standardization.
  - Similarity with past proven designs.
- Update design plans as development proceeds.

REQUIRED DOCUMENTS

1. Design procedure.
2. Documentation of design personnel qualifications.
3. Design plans and other design documents.

46. What design inputs does the Standard suggest, and what form should design outputs take?

CAPSULE ANSWER

Design input is made up of a precise definition of customer needs, plus any applicable laws, regulations, and statutes. Design output is, ideally, a mirror image of design input, expressed in technical language.

Design input is the information a company obtains when it is preparing to carry out a design. Design output is the result of the design process—usually, a set of instructions and information for the production of a product or service.

When gathering design input, the Standard requires that all laws and regulations, results of contract reviews, and statements of customer requirements be taken into account. Computer-aided design (CAD) and computer-aided manufacturing (CAM) systems must be capable of two-way communication with customer systems, except as waived by the customer.

The better the design input gathered, the more accurately the design will reflect customer specifications, and the more the First Facet of Quality will be fulfilled: "Quality that results from the best possible definition of customer needs."

Design output, whatever form it takes (prints, drawings, plans, and so on), must be the mirror image of the design inputs—except that it expresses customer specifications in technical language that is appropriate to and understood within the design process. Design output must be the end product of a process that includes certain specified techniques (see below).

The more the design output reflects the spirit and the letter of the design input, the more the Second Facet of Quality will be fulfilled: "Quality that results from the best possible product design."

TECHNICAL REQUIREMENTS

1. During the *design input* phase, evaluate:
   - Applicable statutory and regulatory requirements.
   - All incomplete, ambiguous, or conflicting requirements; resolve them with those responsible for imposing them.
   - Results of any contract review activities (Question 40).
2. During the *design output* phase, document and express design outputs in terms of requirements that can be identified.

3. Ensure that design output results from a process that includes:
   - Efforts to innovate, optimize, reduce waste, and simplify.
   - Use of design failure mode and effects analyses (DFMEA).
   - Use of feedback from the field and from the production and testing units.
   - Use of geometric dimensioning and tolerancing (GD&T).
4. Ensure that design outputs:
   - Meet design input requirements.
   - Contain or make reference to acceptance criteria.
   - Identify design characteristics that are crucial to safe and proper functioning of the product.
   - Include reviews of design output documents before release.

TECHNICAL GUIDELINES

DESIGN INPUT

- Design process input includes:
  - Performance requirements.
  - Functional requirements.
  - Descriptions.
  - Environmental and safety regulations.
- Design input requirements should be:
  - Defined, so that their achievement can be verified.
  - Reviewed.
  - Analyzed so that all incomplete, ambiguous, or conflicting requirements are resolved.
  - Recorded in a design description document.
- The design input document serves as the definitive up-to-date reference document as the design progresses to completion. It should:
  - Include details, agreed on between the supplier and the customer, on how customer and regulatory requirements (if any) will be met.
  - Identify design aspects, materials, and processes requiring development and analysis.
  - Be prepared in a way that facilitates periodic updates.

DESIGN OUTPUT

- Design process output should be defined and documented. It can consist of:
  - Drawings.
  - Specifications.
  - Instructions.
  - Software.
  - Service procedures.
- Design output should match up with design input requirements.
- Design output must be verified to ensure that it:
  - Meets applicable regulatory requirements.
  - Contains or references acceptance criteria for subsequent phases.
  - Conforms to appropriate development practices and conventions.
  - Identifies characteristics of product that are crucial to safe and proper functioning.

REQUIRED DOCUMENTS

1. Design procedure.
2. Design input records.

47. Are design reviews required by the Standard?

> **CAPSULE ANSWER**
>
> Design reviews are mandatory activities designed to ensure that the design process stays on track toward developing product designs that conform to customer requirements.

Yes. Design reviews are held for the same reasons verification activities are carried out during the production process: to confirm, at critical points in the design process, that the design, as it is emerging, is meeting the requirements of design input (i.e., the customer's needs and expectations). Design reviews are also held to detect problems and to initiate corrective actions.

Every design activity conducts design reviews. Some are formal, many are not. The Standard does not require that all design reviews be formalized and documented. But, to meet the requirements of this clause, the design plans should indicate precisely the junctures at which formal, documented design reviews are conducted.

Design reviews must be recorded, and it is a good idea to keep these records in the design package or file.

TECHNICAL REQUIREMENTS

1. Plan and conduct formal documented reviews of design results at appropriate stages (defined in the design plan or in the procedure).
2. In the design reviews, include representatives of all functions associated with the design, as well as other specialist personnel, as required.
3. Keep records of design reviews.

TECHNICAL GUIDELINES

- Design reviews, which are formal, documented, systematic, and critical reviews of design results, should be carried out after each design phase.
- Reviews should compare the design with the product or service brief, as applicable. Elements reviewed can include items pertaining to:
  - Customer needs and satisfaction.
  - Product specification requirements.
  - Process specification requirements.
- The purpose of documented design progress reviews are:
  - To ensure that resource issues are resolved.
  - To verify effective execution of development plans.
  - To ensure that requirements are met and that specified methods are being correctly carried out.
  - To identify and anticipate problem areas and inadequacies.
- Outcomes of design reviews should be:
  - Determination of the consequences of all known deficiencies, or understanding of the risks of proceeding.
  - Initiation of corrective actions, ensuring that final design and supporting data meet customer requirements.

REQUIRED DOCUMENTS

1. Design procedure.
2. Design plan.
3. Design review records.

48. What are design verifications/validations?

> **CAPSULE ANSWER**
>
> Design verifications and validations are required activities to confirm that designs fully meet customer needs and expectations.

Design verifications and validations are mandatory activities. The Standard requires design verifications to be performed at the end of the design phase— and, sometimes, at intermediate design phases—to confirm that design outputs conform to design inputs.

A company is also required to operate a complete prototype program (except for standard items, or as waived by the customer). Prototype testing must include product life, reliability, and durability.

At the end of the design activity, conduct a design validation to verify the validity of the design.

TECHNICAL REQUIREMENTS

1. Perform design verifications to ensure that design-stage output meets design-stage input requirements.
2. Design verifications must be recorded.
3. Perform design validation to ensure that the product design conforms to the user's needs and/or requirements.
4. Design validation must include a comprehensive prototype program, except for standard items or as waived by the customer. The program must include, as appropriate:
   ■ Product life.
   ■ Reliability.
   ■ Durability.
5. Prototype and testing activities must be monitored to ensure timely completion and conformity with requirements.

6. Chrysler suppliers must perform design validation/production validation at least once per model year on all products, except as described in the Chrysler specifications.

TECHNICAL GUIDELINES

Design verifications:

- Are normally performed on the final product under defined operating conditions.
- May be conducted in earlier stages.
- Should be carried out, as closely as possible, under actual use conditions.
- Should consist of any two of the following:
  - Design review.
  - Qualification tests.
  - Alternative calculations.
  - Comparison with a proven design.
- Can also consist of:
  - Analytical methods (FMEA).
  - Fault tree analysis.
  - Risk assessment.
  - Inspection/test of prototypes or actual production samples.
- Should be tested to a degree that is commensurate with the risks identified in the design plan.
- Should be carried out by competent persons other than those responsible for the design.
- Must include recording and checking the results.
- Multiple validations may be performed if the product has multiple uses.

Design requalification:

- Should be carried out periodically to ensure that the design is still valid.
- Should include review of customer needs and technical specifications in light of:
  - Field experience.
  - New technology.
  - Process modifications.

- Should ensure that the need for design changes, indicated by production or field experience, is captured and reported for analysis.

49. How must "critical" characteristics" be handled?

Each of the Big 3 identifies certain products, and characteristics of those products, as being "critical" in varying degrees or "key" in some sense.

What is "critical"? Classifications include:

> **CAPSULE ANSWER**
>
> These are product characteristics, designated by a customer or by a supplier, that affect safety, compliance with government rules, form, fit, function, and similar concerns.

- Characteristics that have an effect on safety or represent compliance with legal rules.
- Characteristics that affect fit, form, function, customer satisfaction, and so on.

Each of the Big 3 has its own identification scheme (see below). Each level is identified by its own particular symbol. It is essential for a supplier to understand the characteristics applicable to each customer and follow the individual guidelines.

Suppliers are encouraged to identify "special characteristics" that may require careful control during manufacturing, in order to ensure full compliance with customer requirements.

All "critical characteristics," whether designated by the customer or by the company, must be included on the relevant Control Plan (Question 55), along with the means by which the designated characteristics will be controlled.

TECHNICAL REQUIREMENTS

1. All characteristics discussed here must be included in and controlled by the relevant Control Plan (Question 55). The symbols must also appear on other documents, as specified by customers.
2. Chrysler subdivides critical characteristics into the following categories:
   - Safety characteristics. Special manufacturing control is needed for compliance with customer or government vehicle safety rules.

(See "Shields—Critical Characteristics Guidelines" published by Chrysler.) A diamond symbol identifies relevant documents.

- Special characteristics. Specifications are critical to part function and have special significance for quality, reliability, and durability. (See "Diamonds—Critical Characteristics Guidelines" published by Chrysler.) A diamond symbol identifies relevant documents.
- Critical tooling characteristics. Certain characteristics of developmental parts, fixtures, gages, and initial product parts require special control. (See "Pentagon—Critical Verification Symbol Guidelines" published by Chrysler.) A pentagon symbol identifies relevant documents.

3. "Significant characteristics," identified by the supplier, are product/process characteristics that require priority attention and/or control to ensure that overall product performance standards are met. The company develops these through knowledge of the process.

4. Ford defines critical characteristics as product requirements or process factors of Control Item parts that can affect:
- Government rules.
- Safe vehicle/product function.

   An inverted delta symbol identifies and must be used on all relevant documents.

5. Other Ford rules with respect to critical characteristics include:
- Ford design or quality engineers must approve initial and revised Control Plans, Design FMEAs (where relevant) and Process FMEAs (Question 30).
- The inverted delta symbol must precede the Ford part number on shipping containers.

6. For Ford's Control Item Fasteners, the Control Plan entries require the following:
- For heat-treated parts, analysis and test of at least one sample to determine conformance to specifications for chemical composition and quenched hardness. A sample of each additional coil or bundle must be tested for either of those characteristics.
- For non-heat-treated parts, material identification must be visually checked to make sure the mill heat number agrees with the steel supplier's mill analysis document and relevant specifications. Each coil or bundle must be tested for hardness, composition, and so on.
- Lot traceability must be maintained (Question 53).

7. General Motors utilizes "key characteristics" symbols and definitions as follows:

- A diamond symbol indicates a "Fit/Function" (F/F) characteristic—a key characteristic not relevant to safety or legal issues. It designates a product characteristic that may affect customer satisfaction with the fit, function, mounting, or appearance of a product. It also indicates product characteristics that may present special process problems or issues.
- An inverted delta symbol indicates a "Safety/Compliance" (S/C) characteristic—one that could affect product safety or compliance with government regulations.

TECHNICAL GUIDELINES

- For Chrysler:
  - "Shields—Critical Characteristics Guidelines."
  - "Diamonds—Critical Characteristics Guidelines."
  - "Pentagon—Critical Verification Symbol Guidelines."
- For Ford:
  - "Guidelines for Production Parts" (Ford 1750).
- For General Motors:
  - "Key Characteristics Designation System."

REQUIRED DOCUMENTS

1. Detailed procedures that address the requirements discussed here for:
   - Advanced quality planning.
   - Design control.
   - Handling, storage, packaging, preservation, and delivery.
   - Production part approval.

AUDIT ISSUES

The required symbols must be used wherever particular customer(s) have asked that they be used.

50. Must design changes be controlled?

Yes. The Standard requires a company to plan, manage, control, and document all design changes. Customer approval—or waiver—of design changes is required, prior to putting the change into production.

The safest route to use for design change approvals is the same basic system that is used for initial design review and approval activities.

TECHNICAL REQUIREMENTS

1. Authorized personnel must identify, review, approve, and document all design changes before they are put into effect. Before implementation, design changes must have written customer approval or waiver of such approval.

2. For design changes to Ford Control Item parts (Question 49), Ford Product Engineering approval is required, on a specialized form available from Purchasing.

3. The Production Part Approval Process (Question 39) should be utilized where applicable.

4. For proprietary designs, the company must, in conjunction with the customer, evaluate the following characteristics:
   - Form.
   - Fit.
   - Function.
   - Performance.
   - Durability.

TECHNICAL GUIDELINES

Designs of products and services should not be changed without due cause and consideration. The documented design change process should:

- Involve people from all functions affected by the change.
- Control the change of documents that:
  - Define the design baseline.
  - Authorize the work needed to implement design changes that can affect a product during its entire life cycle.

- Include a review to see whether:
  - The changes influence previously approved design verification results.
  - The changes are impacting the whole design in an unanticipated way.
- Provide for:
  - Necessary approvals.
  - Timing for changes.
  - Removal of obsolete drawings and specifications from work areas.
  - Confirmation that changes are made as specified.
- Communicate new design output to all concerned parties, including customers, when design changes will affect product or service characteristics and/or performance.
- Verify that only authorized design changes have been made.
- Provide for emergency changes to prevent the production and delivery of nonconforming products.

REQUIRED DOCUMENTS

1. Design control procedure (addressing design changes).
2. Records of customer approvals or waivers.

Controlling the Process

51. What must a company do to control its process planning?

In a requirement that is only mentioned indirectly in ISO 9001, QS-9000 sets forth rules for facility and process planning. This planning is to be carried out as part of the documented advanced quality planning process (Question 30).

In its general guidelines for plant layouts, the Standard requires that they maximize the efficient flow of material and the value-added use of floor space.

Even more critical, the company must have a process for evaluating the effectiveness of existing operations and processes. Although it is not explicitly stated, this should be a documented process defined by a procedure.

> ### CAPSULE ANSWER
>
> The Standard requires that facility and process planning be carried out on a cross-functional basis, against specific standards, and that process effectiveness be evaluated against defined factors.

TECHNICAL REQUIREMENTS

1. Using a cross-functional team approach, the company must plan facilities, processes, and equipment as part of the advanced quality planning process (Question 30).
2. Plant layouts should:
   - Facilitate synchronous material flow.
   - Maximize value-added use of floor space.
   - Minimize material travel and handling.
3. The company must evaluate the effectiveness of existing operations and processes in light of:
   - Automation.
   - Balance of operators and lines.
   - Ergonomics/human factors.
   - Inventory levels (storage and buffer).
   - Overall work plan.
   - Value-added labor content.

REQUIRED DOCUMENTS

1. Advanced quality planning process procedure.
2. Procedure for evaluating process effectiveness (can be part of above).

AUDIT ISSUES

The *Quality Systems Assessment* (QSA; Question 15) document asks about only:

- The use of cross-functional teams.
- Whether plant layout minimizes material travel and maximizes use of floor space.

A supplier should expect to be asked how these analyses are carried out, and what evidence is available as proof that they are carried out on a regular basis.

52. How should tooling be handled?

QS-9000 specifically requires a tooling management system. Although it does not call for a "documented" system (i.e., a written procedure), it is highly advisable to document the tooling management process by writing up a procedure.

> **CAPSULE ANSWER**
>
> Usually, a supplier must provide the technical resources and system for tooling management, whether it is done internally or by external resources.

A tooling management system can be carried out externally, but there must be a tracking and follow-up system to control it.

If a company uses tooling provided by its customers, the tooling must be visibly and indelibly marked as to its ownership. This is equally true for tooling in which customers own a partial share.

TECHNICAL REQUIREMENTS

1. The company must provide "appropriate technical resources" for the design, fabrication, full dimensional inspection, and management of tools and gages. This system must cover tooling:

- Maintenance, repair facilities, and personnel.
- Storage and recovery.
- Setup.
- Changes (for perishable tools).

2. If these activities are subcontracted, the company must operate a tracking and follow-up system.
3. The company must provide visual and permanent marking as to the ownership of customer-owned tools and equipment.

REQUIRED DOCUMENTS

1. Tooling management procedure.
2. Tooling maintenance records.

53. Does the Standard really require a company to label every single thing in each of its buildings?

No. "Where appropriate" is the term that the Standard uses under its Product Identification and Traceability requirements. That clause requires a company, "where appropriate," to:

> **CAPSULE ANSWER**
>
> To the extent that these activities assist in controlling quality, a company should be able to trace and identify materials and products throughout the production process and in the field as well.

- Identify materials and products at all stages of the process.
- Trace materials and products at all stages of the process, and beyond.

When are "identification and traceability" appropriate? Here are a few suggestions:

- When a company needs to examine all materials supplied by a particular vendor (for example, if it suspects or is informed that a vendor's product is defective).
- When a company needs to recall products shipped to customers, for safety, product liability, or customer satisfaction reasons.
- When identification and/or traceability are required by customer contract.

- When a company needs to isolate specific products during production in order to:
  - Verify the presence of nonconforming conditions.
  - Apply corrective actions.

The Standard provides a great deal of latitude as to when, and under what circumstances, materials or products should be identified and/or traceable. Practically speaking, there are very few situations in which some level of these capabilities would not be expected in an ISO 9000 quality system.

The Standard also requires another form of identification to indicate the "inspection and test status" of a product. Virtually all products get inspected in some form, usually at several points in the process (Question 61). The requirement here is that anyone must be able to tell, by looking at the product:

- Whether is has been inspected.
- Whether it has passed or failed, or its status is unknown.

The Standard says this requirement is to be fulfilled by "suitable means"— usually, some form of tag: a label, hang-tag, traveler, or whatever. It can also be fulfilled by location, as long as the location makes the inspection and test status "inherently obvious." For example, an area of a facility may be designated as a "finished goods" area. All product located there has, by definition, been inspected and passed. However, the Standard specifically says that location in and of itself does not constitute adequate indication of inspection and test status *unless* the status is "inherently obvious."

TECHNICAL REQUIREMENTS

1. In situations where a product's identity is not inherently obvious, it must be identified "by suitable means":
   - From receipt.
   - During all stages of production, delivery, and installation.
2. These activities must be documented with procedures.
3. Where "traceability is a specified requirement" (i.e., by the customer), a company must:
   - Uniquely identify individual products or batches.
   - Record the identification.
   - Document these activities with procedures.

4. Inspection and test status must be identified by suitable means throughout the process, to ensure that only product that has undergone and passed inspection reaches the customer.

5. Product location alone does not meet this requirement unless the location makes the inspection and test status inherently obvious.

TECHNICAL GUIDELINES

■ Traceability can involve high cost; implement a system only to the extent that it meets a well-defined need.
■ Identification is often useful for:
  – Tracking process changes.
  – Tracking performance of personnel, tooling, and so on.
■ Identification can consist of:
  – Marking or tagging a product or container.
  – Color coding (especially for bulk materials, batches, or defined lots).
  – Documentation.
■ For processed materials:
  – Maintain appropriate identification throughout the production process where in-plant material traceability is important to quality.
  – Retain samples from each lot for a defined period of time.

INSPECTION AND TEST STATUS

■ The system for inspection and test status should:
  – Verify that required inspections and tests have been performed.
  – Clearly indicate the results of tests.
■ Status can be indicated via:
  – Marking or tagging.
  – Signage.
  – Physical segregation (the safest alternative).

REQUIRED DOCUMENTS

1. Procedure(s) for identification and traceability (as required).
2. Records of batch identification.

54. What are the requirements for process control?

Element 4.9, Process Control, is the most extensive and prescriptive element of QS-9000. The overall requirement is twofold; production processes must be:

> **CAPSULE ANSWER**
>
> A production process must be planned, and production activities must be carried out under a set of specifically defined controlled conditions.

1. Planned.
2. Carried out under controlled conditions.

A related requirement states that production scheduling must be order-driven. In other words, production scheduling must be traceable to customer orders and tied into delivery lead time requirements, as appropriate.

The balance of the Process Control element describes what the Standard means by "controlled conditions." The high points are:

- Written instructions (controlled under the Document Control system; Question 38) must be posted at all places where production and related work are carried out. These instructions must tell operators how to:
  - Do production-related activities.
  - Monitor the process.
  - Inspect the product.
  - Carry out other quality-related activities.

    These are sometimes called "work instructions" or "Level 3 documents" (Question 37).
- "Workmanship standards"—in written, graphic, or sample form—must be available to employees so they can judge how well they are meeting quality requirements.
- Product characteristics must be monitored to ensure that defined quality standards are being met. This is especially important for "special" or "critical" product characteristics (Question 49).
- The process must be proven to be capable of meeting the requirements set down for it, and monitored on an ongoing basis to ensure that standards are being met. (Addressed in more detail in Question 58.)
- Production-related equipment must be covered by documented preventive maintenance measures. (Addressed in more detail in Question 57.)
- Changes to processes must be controlled and, where required, approved by the customer. (Addressed in more detail in Questions 59 and PPAP.)

TECHNICAL REQUIREMENTS

1. Management must identify and plan every production, installation, and servicing activity that affects quality.
2. Production scheduling must be order-driven.
3. The plans must ensure that the above processes are carried out under "controlled conditions."
4. Controlled conditions include:
   - Documented procedures detailing how production activities are carried out, where the absence of such procedures could adversely affect quality. This is a separate issue from the required "process monitoring and operator instructions" (Question 37).
   - Appropriate approvals of processes and equipment. QS-9000 requires documented planning activities for facilities, equipment, and processes (Question 51).
     - A procedure must define who has authority to approve new processes and equipment, as well as changes to those in progress or in use.
     - In accordance with the Production Part Approval Process (PPAP) requirement (Question 39), customers will often be involved in these approvals.
   - Verification of job setups. In accordance with documented procedures (most likely, work instructions; Question 59), verification must address:
     - Use of suitable equipment, in a suitable working environment.
     - Compliance with all relevant safety and environmental regulations, as well as reference standards, quality plans, and procedures.
     - Monitoring and control of process parameters and product characteristics, especially "special characteristics" (Questions 49 and 60). The means of monitoring and controlling special characteristics are detailed in Control Plans, which emerge from the advanced quality planning process (Question 30).
     - Workmanship standards that are clearly presented in sample, graphic, or procedural form. (Approved production samples are a good way to meet this requirement.)
     - Where applicable, statistical verification, which is related to PPAP (Question 39), is required.
   - Control of process changes. A company must meet the process change requirements of the Production Part Approval Process (PPAP) system (Questions 59 and PPAP) and should maintain records of the effective dates of process changes.

- Proper conditions for appearance items. Customers will identify any parts being made for them as "appearance items." This situation requires:
  - Appropriate lighting in areas where such parts are checked.
  - Appropriate masters (calibrated; Question 62) for color, grain, texture, and so on.
  - Trained and qualified personnel for evaluating the appearance items.
- Suitable equipment maintenance, including preventive maintenance (Question 57).
- Control of "special processes" (Question 60).
- Process monitoring and operator instructions (Question 37).
- Process capability studies (Question 58).
- For all employees carrying out process and/or monitoring activities, written instructions available at the work sites (Question 37).

TECHNICAL GUIDELINES

GENERAL

- In developing process control systems, consider the following factors:
  - Production process complexity.
  - Availability of proven production processes.
  - Need to develop new processes.
  - Number and variety of processes.
  - Impact of various processes on product performance.
  - Need for various levels of process control.
  - Ability to control equipment consistently.
  - Stability of essential materials.
- Define processes in terms of attributes that directly affect product or service performance, whether observable by the customer or not.
- Ensure an ability to evaluate both types of attributes against defined acceptability standards, on either a quantitative (measurable) or qualitative (comparable) basis.
- Vary the level and intensity of control according to:
  - The effect a given product attribute has on quality; attributes most critical to quality should be identified and kept under closest control.
  - The greater impact of nonconformities downstream.
  - The ability to measure the attribute accurately.
- Use either written or electronic documentation.

RECOMMENDED ELEMENTS OF EFFECTIVE PROCESS CONTROL

- Carefully plan processes that are important to quality.
- Where possible, schedule production in small lots with a goal of a one-piece flow in a synchronous manner.
- Establish an early warning system to identify obstacles to stable production.
- Give special attention to product attributes that are difficult to measure.
- Establish appropriate controls to cover:
  - Materials.
  - Installation and service equipment.
  - Documented procedures.
  - Quality plans.
- Plan and implement effective methods for controlling incoming, in-process, and final product through delivery and until put into service.
- Control—and periodically verify, when important to product quality—auxiliary process aspects such as suppliers, utilities, and environment.
- Control process changes carefully:
  - Clearly designate those responsible and, where necessary, get customer approval (Question 39).
  - Document all changes and cover them with defined procedures.
  - Evaluate the product after any change, to verify that the change has had the desired effect.
  - Encourage efforts to develop new methods for improving process quality.

VERIFICATION

- To minimize the effects of errors and to maximize yields, verify quality status at important points in the production sequence. Verification can be facilitated with:
  - Control charts.
  - Statistical sampling procedures.
- Relate process monitoring directly to finished product specifications, or to an internal requirement:
  - Communicate and document this relationship to all personnel concerned.

- Establish control to ensure that these attributes remain within specifications, or that appropriate adjustments are made.
■ Plan and specify all in-process and final verifications:
- Maintain documented testing and inspection procedures for each quality attribute to be checked.
- Specify equipment, requirements, and workmanship criteria for all verification activities.
- Identify product verification status by stamps, tags, notations, and other suitable means.
■ Maintain differentiation among unverified, conforming, or nonconforming product; identify who did the verification.

PROTECTION

To maintain the suitability of in-process products, provide appropriate storage, segregation, handling, and preservation.

DOCUMENTATION

■ Control process operations to the necessary extent via documented work instructions. Procedures and/or work instructions should describe the criteria for determining:
- Satisfactory work completion.
- Conformity to specifications.
- Conformity to standards of good workmanship.
■ Stipulate the criteria for workmanship, in the clearest practical manner, by using:
- Written standards.
- Photographs.
- Illustrations.
- Samples.
■ Control process documentation as specified in the quality system.

TRACEABILITY

■ Where traceability is important, maintain appropriate identification through all process stages. Markings should be:

– Legible.
– Durable.
– In accordance with specifications.
– Adequate to identify a particular product in event of recall or need for special inspection.

POST-PROCESS ACTIVITIES

Establish and maintain appropriate methods of cleaning, preserving, and packing.

VERIFICATION OF JOB SETUPS

■ Produce enough product for a standard statistical process control (SPC) sample.
■ Enter measurements on a control chart.
■ If data fall within the central third of a controlled limit zone, approve the setup for production.
■ If data fall outside the zone, measure and plot a second subgroup.
■ If these data also fall outside, adjust the setup and run the verification process again.

REQUIRED DOCUMENTS

1. Process control procedure.
2. Work instructions:
   ■ General (as needed).
   ■ Process monitoring and operator instructions (required).

AUDIT ISSUES

Registration assessors practically live in this set of requirements for a major portion of an audit. Count on a detailed examination of:

- Process planning and approval documents.
- Operator instructions. (Do they exist? Do employees know where they are? Are they controlled?)
- Workmanship standards.
- Adherence to PPAP requirements (Question 39).

55. What is a control plan?

A control plan describes the process for controlling how a part, or family of parts, is made. It provides specific instructions not only for making the part, but also for inspecting it and controlling its special characteristics. The control plan is almost always found on the manufacturing floor at the location where the operation is carried out. Operators are expected to understand and follow what the control plan says.

CAPSULE ANSWER

Control plans define how a part is made and how its characteristics, especially "special characteristics," are controlled so that the part meets customer requirements.

The control plan is the main output of the advanced quality planning (AQP) process (Question 30). Under most circumstances, customers must approve control plans as part of the Production Part Approval Process (PPAP) system (Question 39).

Each control plan is subject to change, based on process changes/improvements. It is a living document that must be controlled under the QS-9000 document and data control system.

The typical control plan includes part name and description, key contact name information, and key approvals (e.g., customer). It lists each operation needed to make the part and, for each operation, specifies:

- Machine or device used.
- Characteristics to be controlled (including special characteristics of the product and the process).
- Specifications and tolerances for the characteristics.
- Evaluation or measurement technique.
- Required sample size and frequency.
- Control method.
- Reaction plan (when results are out of specification).

TECHNICAL REQUIREMENTS

1. Control plans must be developed to control the production of all parts and processes. They must be created at the following levels, as appropriate to the planned activity:
   - System.
   - Subsystem.
   - Component.
   - Material.
2. Control plans cover the following phases, as appropriate:
   - Prototype (whether customer requires it or not).
   - Prelaunch.
   - Production.
3. Control plans define the following, at a minimum:
   - Specifications and tolerances.
   - Inspection and measurement techniques.
   - Samples sizes and frequencies.
   - Control methods (defining how the operation is controlled).
   - Reaction plan (corrective actions taken when nonconformities appear).
4. Control plans must be approved by customers, unless approval is waived.
5. Control plans must be living documents that are subject to review, update, and approval when the product changes or when processes:
   - Change.
   - Become unstable.
   - Become noncapable.

56. Does the Standard expect compliance with government safety and environmental regulations?

Yes. Most companies would comply even without QS-9000 making a point of it (by virtue of EPA, OSHA, and other alphabet agencies), but QS-9000 makes several references to compliance with governmental safety, health, and environmental regulations.

> **CAPSULE ANSWER**
>
> A company must document and enforce its policies for health, safety, and the handling of hazardous or toxic substances.

Specifically, a company's business plan (required) must address, as applicable, its intent to deal with these issues. Regulations must also be taken into account in any process for purchasing products that may include hazardous substances. There must also be a process for the proper disposal or recycling of hazardous or toxic substances that may be generated in the purchasing or manufacturing processes. These activities must be documented.

TECHNICAL REQUIREMENTS

1. The required business plan (Question 24) must include and address, as applicable, health, safety, and environmental issues.
2. With respect to purchased products and services and the manufacturing process, the company must have a process to ensure compliance with governmental and safety rules regarding restricted, toxic, and hazardous substances.
3. The company must also have a process for handling, recycling, eliminating, or disposing of hazardous materials. This process should be confirmed by appropriate certificates or letters of compliance.

TECHNICAL GUIDELINES

■ When dealing with suppliers, it is not enough simply to put "Comply with governmental and safety rules with respect to restricted substances" on the purchase order. A process to ensure compliance is necessary.
■ A company's process for dealing with subcontractors should include one of the following:
  – On-site visits to confirm compliance with governmental and safety rules on restricted substances.
  – Supplier evidence of compliance (certificates, and similar proof).

The QS-9000 references provide little guidance on product safety. However, several of the ISO 9000 guidance documents address this issue, as summarized here:

■ The company should:
  – Establish a company policy on safety.
  – Identify relevant safety standards to make formulation of product or service specifications more effective.

- Identify applicable worldwide and local laws and regulatory requirements on safety, to enable those concerned to conduct risk and compatibility assessments.
- Carry out and document design evaluation tests and prototype (or model) tests for safety.
- Develop means of treaceability to facilitate product recall and to allow planned investigation of products and services suspected of having unsafe features.
- Make intended use clear, and issue warnings with respect to known hazardous materials by means of labeling, instructions, and/or promotional media.
- To minimize misinterpretation, especially regarding intended use and known hazards, analyze user instructions and warnings, maintenance manuals, labeling, and promotional media.
- Consider developing an emergency plan in case recall of a product becomes necessary.
■ The product or service design process should take into account:
  - Product safety requirements and regulations, including elements of the company's quality policy that may go beyond existing statutory requirements.
  - Risk of occurrence of failure and the anticipated consequences.
■ The service delivery specification (developed as part of the design process) should take safety requirements into account.
■ Supervisors and workers should be thoroughly trained and proven to understand how their duties relate to safety in the workplace.
■ The significance of quality problems should be evaluated in terms of product safety.
■ Product safety should be taken into account when disposing of nonconforming materials, taking remedial and/or corrective actions, and making product recall decisions.

REQUIRED DOCUMENTS

1. Separate procedures may or may not be warranted for compliance with:
   ■ Governmental safety requirements.
   ■ Governmental environmental requirements.
   ■ Handling and/or recycling of hazardous substances.

However, the issues raised in the requirements must be addressed in procedures.

2. Business plan.

AUDIT ISSUES

Usually, companies that are affected by governmental safety and environmental issues already have processes in place to address them. These measures, and related documentation, are considered relevant by the QS-9000 system and need to be addressed. Environmental matters may be assessed as part of the audit of purchasing or as part of process control.

57. Does the Standard require preventive maintenance of production equipment?

CAPSULE ANSWER
Key process equipment must be covered by an effective, planned, total preventive maintenance system.

Yes, and the requirement is very specific and very strict. This is bad news for companies that "have no time" for preventive maintenance, but have plenty of time for their equipment to be down, waiting for repairs.

The Standard requires that production equipment must be maintained. Most companies already do that. But the Standard also requires a complete system for preventive maintenance of what it calls "key process equipment." It is up to each company to determine what that is. Judgment calls will be checked by the assessors.

Preventive maintenance means more than just squirting oil on moving parts every so often. It involves:

■ Setting schedules for machine checks and adjustments, to ensure that the machine's capability remains as expected.
■ Adjusting preventive maintenance frequencies and activities in response to actual experience, such as:
  – Repair activity.
  – Unscheduled downtime.
  – Inspection results.
  – Analysis of statistical data.

- Maintaining an inventory of essential repair parts. Actually, an "inventory" per se is not required. But a company must show that it has ready access to such resources. In most companies, that means an inventory of some sort—and some control over it!

Tooling, including perishable tooling, must also be maintained on a scheduled basis (Question 52).

The preventive maintenance requirement presents another chance for "preventive action." A company that has what the Standard calls an "effective, planned, total preventive maintenance system" is keeping its process at its peak of capability, maintaining maximum uptime, and, most likely, saving itself a lot of money as well. Preventive maintenance is not just a paperwork exercise. It is a genuine opportunity for improvement.

TECHNICAL REQUIREMENTS

1. The company must:
   - Identify key process equipment.
   - Provide appropriate resources for machine/equipment maintenance.
   - Develop and implement an effective, planned, total preventive maintenance system.
2. Minimum requirements of the preventive maintenance system include:
   - A documented procedure.
   - Scheduled maintenance activities.
   - Availability (inventory) of replacement parts for key manufacturing equipment.
   - Predictive maintenance methods that take into account:
     - Correlation of statistical process control (SPC) data to preventive maintenance activities.
     - Fluid analysis.
     - Important characteristics of perishable tooling.
     - Infrared monitoring of circuits.
     - Manufacturer's recommendations.
     - Monitoring of uptime.
     - Tool wear.
     - Vibration analysis.
3. Tooling must also be maintained (Question 52).

TECHNICAL GUIDELINES

- Perishable tooling is included in the requirement. The company must use statistical methods to establish predictive maintenance schedules for such tools. Last-piece inspection, for the purpose of predictive maintenance, is not enough.
- Including equipment repairs in the maintenance records is an excellent idea. Preventive maintenance schedules can be adjusted to reflect repair activity, as appropriate.

REQUIRED DOCUMENTS

1. Equipment maintenance procedure.
2. Maintenance records.

AUDIT ISSUES

- Much attention is devoted to equipment maintenance schedules and to records confirming that the maintenance was carried out.
- Assessors expect to find that preventive maintenance schedules have been adjusted to reflect quality-related occurrences such as downtime, equipment wear, repair activity, and so on.
- Assessors also want to see that maintenance activity has been taken into account when checking process capability (Question 58).

58. What does the Standard mean by "process capability"?

A company must prove to itself (and, more important, to its customers) that its process is capable of meeting customers' quality requirements. It must do this:

- Before beginning production, as part of the Production Part Approval Process (PPAP) (Question 39).
- On an ongoing basis after production begins.

> **CAPSULE·ANSWER**
>
> Process capability is a statistical method designed to assess how well an overall process meets the quality requirements for "special characteristics" identified by the company and/or the customer.

That very general statement applies to all characteristics of the product. But the requirement focuses specifically on what the Standard calls "special characteristics" (Question 49). These are characteristics, identified by either the customer or the supplier, that affect:

▪ Safety.
▪ Part function.
▪ Compliance with customer and/or governmental vehicle safety, emission, or noise requirements.

These characteristics are identified during advanced quality planning (Question 30) and included on the relevant Control Plan.

Before beginning a new process, a company must run a process capability study (Ppk). This is a statistical analysis, applied to special characteristics identified from a sample run of parts, to see how well the process is meeting the quality requirement for a particular special characteristic. Using statistical process control charts, data are gathered, analyzed, and reduced to a value. The study and the resulting value are submitted to the customer as part of the Production Part Approval Process (PPAP) (Question 39). They must meet customer requirements before production can commence. If the value does not meet customer requirements, the company must take corrective action on the process until it does.

After production part approval is obtained, ongoing process performance studies must be run. They apply most specifically to the special characteristics called out on the Control Plan. As a rule, statistical process control is also used. Data on special characteristics are recorded, tracked, and analyzed for stability. Control charting can result in corrective action, when the process becomes unstable.

Data are also reduced to a value in order to determine the "capability" of the process. If this value does not meet or exceed the minimum set by the customer, corrective action on the process must be taken as specified in the Control Plan and/or, as relevant, by the customer.

The key to meeting these requirements is to understand a customer's requirements for stability and capability and work closely with the customer to make sure the process meets those requirements.

TECHNICAL REQUIREMENTS

1. The company must carry out process capability studies covering all special characteristics (Question 49) identified by the customer or the company. These include:
   - Preliminary process capability studies (Ppk).
   - Ongoing process performance evaluation (Cpk).
2. The results of the preliminary process capability studies should be reviewed with the customer. They must meet customer requirements. If the customer states no requirement, preliminary results (less than 30 days of activity) of Ppk value greater than or equal to 1.67 are considered acceptable.
3. When the results of preliminary studies are unacceptable, mistake-proofing and/or other continuous improvement methodologies must be implemented.
4. To determine ongoing process performance, the company must carry out ongoing process capability studies for all defined special characteristics, where variable data are available. The results of these studies must meet customer requirements.
5. Where customers do not specify a requirement, the following values apply:
   - For stable processes/normally distributed data: Cpk greater than or equal to 1.33.
   - For unstable processes, where output meets specifications in a predictable pattern, Cpk greater than or equal to 1.67.
6. When characteristics identified on the Control Plan become either unstable or noncapable, the reaction plan specified on the Control Plan must be initiated. This usually consists of:
   - Obtaining customer review and approval where required.
   - Including segregation of affected items, and 100 percent inspection, in the reaction plan.
   - Implementing a defined corrective action plan that states assigned responsibilities and specific timing.
7. Other ongoing process performance requirements include:
   - Noting key process events, such as machine repair and tool change, on control charts.
   - Achieving continuous improvement, regardless of the demonstrated process capability or the level of capability required. This applies to all characteristics, but most especially to special characteristics.

TECHNICAL GUIDELINES

- Preliminary studies are short-term only. They must not be used to predict, on a long-term basis, the way variation in people, methods, equipment, and other factors, will affect process performance.
- Measurement system analysis (Question 64) should be performed on measurement systems referenced in the Control Plan, to assess the impact of measurement error on process capability results.
- Short-term studies should be based on a minimum of 25 subgroups containing at least 100 individual readings.
- Processes should be verified as being capable of producing in accordance with product specifications. Establish control to:
  - Ensure that attributes remain within specification.
  - Trigger necessary changes.
- Process verification should cover:
  - Material.
  - Equipment.
  - Computer systems/software.
  - Personnel.
- Equipment to be tested and proven for accuracy includes:
  - Fixed machines.
  - Jigs.
  - Tooling and related processes.
  - Computers used to control processes.
  - Associated software.

REQUIRED DOCUMENTS

1. Procedure(s) for carrying out capability studies.
2. Records of capability studies.

AUDIT ISSUES

- Meeting this requirement creates a wealth of records and data for evaluation by assessors. Proper maintenance of these documents is essential.
- It is important to establish adequate linkage among the elements involved in generating capability studies:

- Special characteristics.
- Customer requirements.
- Control plans.
- Measurement systems.

59. What other requirements are there for process control?

The Standard specifies some additional requirements involving job setups and process changes. These apply to all suppliers. Another requirement applies only to companies that make parts designated as "appearance items" by their customers.

> **CAPSULE ANSWER**
>
> The Standard requires a company to verify job setups, get customer approval for process changes, and use appropriate equipment and personnel for evaluating "appearance items."

Job setups must be verified. Before full production begins on a particular job setup, initial parts must be checked (especially their special characteristics; Question 49) to make sure that they are consistently within specifications. Most companies use statistical process control (SPC) for verification (see Technical Guidelines, page 161). Ford, in fact, requires SPC. Whatever method is used, it must be covered by written instructions and documented with records.

The next requirement is, potentially, more burdensome. According to Production Part Approval Process (PPAP) requirements, customers must approve key elements of a manufacturing process before full production can begin (Question 39). Even after this has been done, the Standard requires that any change in the following process elements must be approved by the customer (unless approval is waived by the customer):

- Part number.
- Engineering change level.
- Manufacturing location.
- Material source(s).
- Production process environment.

If a decision is made to move a machine 10 feet, in a process already approved by a customer via PPAP, the company must at least inform the customer of the move. Depending on the severity of a change, the customer may require another PPAP submission. All this notwithstanding, the

Standard encourages companies to make process changes (with appropriate approvals, of course) in the spirit of continuous improvement.

The Standard has special requirements for companies that make "appearance items"—parts that have significant cosmetic criteria, as defined by customers. A company that makes these kinds of parts must:

- Utilize an appropriate environment for evaluating them.
- Maintain (and calibrate!) the masters used for verifying color, grain, and texture.
- Assign employees who have documented qualifications to make appearance evaluations.

TECHNICAL REQUIREMENTS

1. Job setups must be verified to ensure that the process is producing conforming parts in accordance with written instructions:
   - The job setup confirmation must be recorded.
   - Comparisons of last-off parts with setup data are recommended.
   - Ford requires job setups to be verified via statistical analysis, using statistical process control (SPC; see Technical Guidelines, page 161) for all "critical" and "significant" characteristics.
2. Where customers require production part approval (Question 39), the approval covers a specific product, process, and manufacturing environment. Customer approval must be obtained (except where specifically waived by customers) whenever there are changes in:
   - Part number.
   - Engineering change level.
   - Manufacturing location.
   - Set of material sources.
   - Production process environment.
3. The company must keep a record of the dates of process changes.
4. Companies that make parts designated by their customers as "appearance items" must use:
   - Evaluation areas that have appropriate lighting.
   - Masters, as appropriate, for color, grain, and texture.
   - Proper maintenance of appearance masters and other evaluation equipment.
5. Personnel who have documented qualifications to make appearance evaluations.

TECHNICAL GUIDELINES

■ This is the recommended approach for verifying job setups using SPC:
  – Make enough product to constitute an SPC subgroup.
  – Take readings on special characteristic(s) and enter these on the control chart.
  – If all readings fall within the central third of the control limit zone, approve the job for production. If they do not, make another subgroup, take more readings, and enter them again.
  – If the new results also fail, adjust the setup and repeat the process.
  – If the new results fall within the central third of the zone, approve the job for production.

■ Although changes in the process must be approved by the customer (as described above), nothing here is meant to discourage process changes in pursuit of continuous improvement (Question 27).

■ Customers must be informed of all changes in a process environment (no matter how seemingly trivial). A company should obtain direction from customers as to whether full PPAP submission is necessary in such cases.

REQUIRED DOCUMENTS

1. Job setup instructions.
2. Records of verification of job setups.
3. Production Part Approval Process (PPAP) procedure.
4. PPAP records.
5. Records of qualifications of production employees, including those doing product appearance evaluations.

AUDIT ISSUES

A company that makes "appearance items" as designated by a customer, and uses masters for color, grain, texture, and other characteristics, must make sure the masters are maintained as required and are covered under the company's calibration system (Question 62).

60. The Standard mentions "special processes." What are they, and why do they deserve their own question?

"Special processes" are difficult to explain, but not that hard to understand once you think about it.

A "special process" makes something that cannot be fully and independently checked for quality at the time the activity is carried out:

- With some special processes, the quality of the work cannot be checked at all without destroying the item or otherwise making it unfit for use. A classic example is a weld. The only way to tell whether a weld is good is to break it or cut it.
- For some products, the quality of the workmanship cannot be judged until later in the process or later in the product's life, long after it is delivered to the customer.

Because special processes do not lend themselves to "traditional" inspection means, the Standard requires that they are treated:

- By specifically qualified and trained people.
- Under conditions of strict monitoring and control.
- In accordance with "specified" (i.e., written) requirements.

Usually, clear work instructions will spell out all critical process steps. Specific and intensive training for operators is also common.

TECHNICAL REQUIREMENTS

1. Special processes must be carried out:
   - By qualified operators.
   - Under continuous monitoring and control of process parameters.
   - In accordance with specified requirements for qualification of process operations, equipment, and personnel.
2. Records must be kept of qualified processes, equipment, and personnel.

TECHNICAL GUIDELINES

- Here are some typical special processes situations:
  - Quality characteristics do not emerge until later in the manufacturing process.
  - Measurement is impossible or destructive.
  - Results within the process cannot be measured until much later in the life of the product. Examples are metals (strength, fatigue life, corrosion resistance), plastics (dyeability, shrinkage), and documents (correctness). (Special processes are most common among processed materials.)
- Requirements for monitoring special processes are:
  - Tight measurement assurance.
  - Perfect calibration.
  - Exceptional statistical techniques.
  - Employees who have exceptional skills and abilities.
  - More frequent verifications.
  - Verification techniques for pressure, time, temperature, flow, environment, and measurement levels.
  - Certification records on personnel, processes, and equipment, as appropriate.
- Verifications must monitor:
  - Accuracy and variability of equipment used to make and/or measure a product, including settings and adjustments.
  - Skill, capability, and knowledge of operators who must meet the quality requirements.
- Upon detection of conformity failure, the action taken to correct the process may involve temporary suspension of the process until the causes are identified and eliminated.

REQUIRED DOCUMENTS

1. Work instructions.
2. Records on special processes.

AUDIT ISSUES

Assessors are particularly interested in verifying that operators involved in special processes have the required training and qualifications.

Verifying Conformance to Requirements

61. What kinds of measurements and tests are companies required to carry out?

CAPSULE ANSWER

The Standard requires a company to verify, at critical points throughout the process, that products meet specified requirements. This can be carried out via inspections, among other means.

QS-9000 specifically says that process activities should focus on defect prevention rather than defect detection. Many elements of the required system, from Advanced Product Quality Planning (Question 30) forward, aim the system at prevention (preventive action) rather than detection.

However, detection is still important because the overall goal of the system is to ensure that any product that reaches the customer will meet the customer's requirements.

The Standard calls for three levels of inspection and testing, documented in procedures and in related documents such as control plans:

1. Receiving inspection, to verify that incoming material meets requirements. (Keep in mind that the objective can be met with means other than receiving inspection.)
2. In-process inspection, to prevent nonconforming product from proceeding further, thereby consuming resources and imperiling customer satisfaction.
3. Final inspection, to verify that all other inspections have been satisfactorily completed.

Records must be kept of inspections. As a practical matter, a system may call for self-verification by operators as they perform work, and not every "informal" inspection needs to be recorded. A company must, however, identify key inspection points in the quality plans (control plans) and keep records of:

- Inspection activities.
- The authority (person responsible) for releasing product.

TECHNICAL REQUIREMENTS

1. The Standard requires verification that materials and products meet specified product requirements.
2. All process activities should focus on defect prevention rather than defect detection.
3. A company must establish and maintain documented procedures for verifying product conformance and for recording the results in a quality plan (control plan) or in procedures.
4. For attribute sampling plans, acceptance criteria must be zero defects. Other acceptance criteria must be written down and approved by the customer.
5. When required by a customer, a company must use accredited laboratory facilities.
6. The quality plan must include layout inspections and functional tests at customer-established frequencies. Results must be made available for customer review when requested.
7. Nonconforming material procedures must be applied (Question 65) if any product fails to pass inspection.

RECEIVING

8. The company's system must prevent incoming material from being used or processed until it has been proven to meet defined requirements, as set forth in quality plans or procedures.
9. Quality plans or procedures must therefore call for one or more of the following methods:
   - Receiving inspection/testing.
   - Product evaluation by accredited contractors or test laboratories.
   - Evidence from a supplier of its own control measures, to ensure that the requirements have been met for:
     - Statistical data.
     - Audits (by the customer or by a third party) of a supplier's location(s).

– Warrants or certifications from the supplier, especially covering test results.

10. If it is necessary to release an incoming product without verification, the product must be identified and recorded so that it can be traced and recalled if nonconformities appear later.

IN-PROCESS INSPECTION AND TESTING

11. Inspections and/or tests must be defined in the quality plan (control plan) and/or documented procedures.
12. A product must be held until required inspections have been carried out and documented, except when traceability and positive measures are in force.

FINAL INSPECTION AND TESTING

13. Final inspections and tests must be carried out as written in a company's quality plan (control plan) and/or documented procedures.
14. Final inspections must include verification that all required inspections or tests (receiving, in process, and so on) have been satisfactorily completed.
15. A product must not be released until all verification measures specified by the quality plan (control plan) have been completed, documented, and authorized.
16. As set forth in the quality plan and/or procedures, there must be complete evidence that a finished product conforms to specified requirements. A company should not ship any product until all activities outlined in the quality plan and/or documented procedures have been completed and recorded.
17. Records of inspection and testing activities must be maintained and should show:
    ■ That a product has undergone all necessary inspections.
    ■ Defined acceptance criteria.
    ■ The authority responsible for the product's release.

TECHNICAL GUIDELINES

RECEIVING INSPECTION

- Inspection is just one method of meeting the measurement and testing requirement. The Standard does not require inspection per se, when other procedures can deliver the required level of confidence.
- The precise method to be used depends on:
  - The importance of the item to the quality of the finished product.
  - The state of control under which the item was produced.
  - Information provided by the vendor.
  - Cost impact.
- To protect incoming materials from premature consumption, they should be segregated and/or marked.
- Incoming product should not be released without verification, unless it is still possible to:
  - Evaluate products objectively.
  - Remedy detected nonconformities without negatively affecting adjacent, attached, or incorporated items.
  - Define an authority for early release, and implement traceability measures for the products involved.

IN-PROCESS INSPECTION AND TESTING

- Conduct these activities with all forms of products, to allow early recognition of nonconformities and effective corrective actions.
- Do verifications at a location that is as close as possible to the point of creation of the inspected attribute. Systems can include:
  - Setup and first-piece inspection.
  - Inspection and testing by the machine operator.
  - Automatic inspection or testing.
  - Fixed inspection stations throughout the process.
  - Monitoring of specific operations by patrolling inspectors.
- When the people responsible for producing the product or service are also doing the inspections and tests, ensure that the quality plan or associated procedures provide for objectivity of results.

FINAL INSPECTION AND TESTING

- Ensure that the plan and/or procedures for final inspection and testing address all specified product release characteristics.
- Verifications can consist of either acceptance inspections and tests, or periodic inspections and tests.
- Acceptance inspections or tests include:
  - 100 percent inspection.
  - Lot sampling.
  - Continuous sampling.
  - Comparison with purchase order.
  - Continuous or periodic product quality auditing of sample units selected as representative of completed lots.

RECORDS

Records should definitively prove that inspected and tested products have fulfilled quality requirements, and should indicate any relevant regulatory and/or product liability requirements.

SPECIAL CONSIDERATIONS FOR PROCESSED MATERIALS

To avoid the possibility of cross-contamination (intermixing of materials), and in cases where raw material goes directly into the process without evaluation (i.e., pipeline shipments), it is essential to develop a relationship of mutual confidence with the vendor.

REQUIRED DOCUMENTS

1. Procedures covering inspection and testing.
2. Records of inspection and testing.

AUDIT ISSUES

- Do the people responsible for inspections and tests have the necessary qualifications and authority?
- Are acceptance criteria clearly defined?
- Are actual inspection and testing activities in conformity with the quality (control) plan?

62. How are companies to handle measurement systems and equipment?

> **CAPSULE ANSWER**
>
> Measurement systems and equipment must be appropriate to the task, accurate enough to create confidence in results, and regularly maintained and calibrated under documented procedures.

Although most companies striving for QS-9000 registration have operated "gage calibration" systems for years, this is still one of the toughest requirements to meet. It is strict, specific, and very easy to audit (probably one of the reasons why it draws a high level of attention from registration assessors).

Another fact is, sadly, a bit of a "dirty little secret." Some companies have historically maintained a paperwork "facade" of a gage calibration system and have gotten away with it because supplier audits in years gone by focused on paperwork only. But QS-9000 assessors go much deeper. If a company is trying to get away with a "sticker and snicker" system, it will take the auditors about 15 minutes to find it. So the time to get "religion" on gage calibration is now.

What exactly is required here?

All devices used to inspect, measure, or test product and product-related characteristics must be controlled. Although the requirement specifies devices used "to demonstrate the conformance of product"—which would technically exclude devices used in non-product-related areas such as tool building, equipment maintenance, and so on—ALL measuring devices should be controlled and calibrated. It a measurement is worth making, it is worth making with an instrument in which a company can have confidence. Only a well-designed calibration system can give that kind of reliability.

The elements of control—spelled out in a documented procedure, of course—are:

- All devices must be identified:
  - By a unique name or number.
  - With an indicator, or an approved record of their calibration status on the date last checked. The date they are next due to be checked should be part of the record.
- All devices must be capable of the measurement precision expected of them.
- All devices must be checked (compared), at defined intervals, with devices whose calibration history is traceable to national or international standards. (This generally means having an external calibration service calibrate master devices, against which the company calibrates its own devices.)
- Calibration methods must be defined and (preferably) documented.
- The record must show the initial reading of each device, the condition of the device, and the amount of adjustment or variation (if any) needed to bring it back into spec.
- Procedures must cover the appropriate handling of devices to preserve their measurement integrity.
- Customers must be notified when calibration activity suggests that a suspect product has been shipped.
- Statistical analysis of all measurement systems must be carried out (Question 64).

TECHNICAL REQUIREMENTS

1. A company must maintain documented procedures to control, calibrate, and maintain all inspection, measuring, and test equipment (including software) used to assess and demonstrate the conformance of products to specific requirements.
2. A company must identify any measurement uncertainty and ensure that the measurement uncertainty is consistent with the required measurement capability.
3. A company must establish the extent and frequency of its instrument calibration, and maintain records as evidence of control.
4. After determining the measurements to be made and the accuracy required, a company must:

- Select inspection, measuring, and test equipment that is capable of the accuracy and precision needed.
- Calibrate and/or adjust all equipment that can affect product quality:
  - At prescribed intervals, or prior to use.
  - Against certified equipment having a known valid relationship to an internationally or nationally recognized standard.
- Where no such standards exist, the company must document the basis for its calibration.
- Define the calibration process, giving all details.
- Identify all measurement equipment with a suitable indicator or approved identification record that shows the calibration status.
- Ensure that environmental conditions are suitable for equipment adjustment and use.
- Handle, preserve, and store measurement equipment so as to ensure accuracy and fitness for use.
- Safeguard measurement equipment from unauthorized adjustments.
- Define the action to be taken when measurement equipment fails.
- Upon determining that measurement equipment is out of calibration, assess and document the validity of previous inspection and test results.

5. A company must maintain calibration records that show:
   - The physical condition, and actual readings, of gages as received for checking.
   - Revisions after engineering changes, where appropriate.
   - Notifications sent to customers when suspect material has been shipped.

6. A company must conduct statistical studies to analyze variations present in each type of measuring equipment system (all systems referenced in the Control Plan) (Question 64).

TECHNICAL GUIDELINES

- Maintain control over all measuring systems used in:
  - Development.
  - Production.
  - Installation.
  - Service.

- ■ This control is intended to provide:
  - – Confidence in any decisions or actions based on measurement data.
  - – Assurance that measurement uncertainty is known and is consistent with the required measurement capability.
- ■ Extend control to vendors' measuring instruments, as appropriate.
- ■ Include in systems and equipment any devices that can affect specified attributes of a product or process, such as:
  - – Gages and similar instruments.
  - – Sensors.
  - – Jigs.
  - – Fixtures.
  - – Comparative references and related guides.
- ■ Establish documented procedures to monitor the measurement process and maintain it in a state of statistical control. Elements of control can include:
  - – Selection of suitable attributes, including range, accuracy, and precision.
  - – Calibration prior to first use.
  - – Periodic recall for adjustment, repair, and recalibration.
  - – Documentary evidence of calibration history, specifying a unique identification for each measuring instrument.
  - – Traceability to reference standards—preferably those with national and/or international recognition—of known accuracy and stability. Alternatively, document the calibration basis.
- ■ When a measuring process is found to be out of control, or measuring and testing equipment is found to be out of calibration, take these corrective actions:
  - – Retest the completed work to determine whether inaccurate measurements were made.
  - – Determine the need for reprocessing, retesting, recalibration, or complete rejection.
  - – Investigate the cause, and implement a solution that will prevent recurrence.
- ■ When the quality of a product or process depends heavily on an ability to measure accurately, obtain more detailed guidance and implementation assistance in ISO 10012-1: *Quality Assurance for Measurement*.

SPECIAL CONSIDERATIONS FOR PROCESSED MATERIALS

- Include the following factors in developing and operating the measurement system for processed materials processing:
  - Correct specification and acquisition of equipment.
  - Initial calibration to validate required accuracy.
  - Periodic recall for adjustment, repair, and recalibration.
  - Documentary evidence of instrument identification, frequency of recalibration, and calibration status.
  - Procedures for recall, handling, storage, adjustment, repair, calibration, installation, and use.
  - Possible use of SPC to maintain processes in statistical control.
- Keep records as documentary evidence of control.

REQUIRED DOCUMENTS

1. Gage calibration procedure.
2. Gage calibration work instructions (almost always).
3. Gage calibration records.

AUDIT ISSUES

Of the many pitfalls inherent in this requirement, the most glaring include:

- Uncalibrated employee-owned gages.
- Out-of-service gages kept where they *might* be used to measure product characteristics.
- Unverified instruments used to measure quality characteristics (rulers, scales, tape measures, and so on).
- Unconfirmed checking fixtures.
- Lack of a consistent and disciplined gage indicator system.
- Lack of operator training as to gage calibration requirements.

63. How can companies reduce
the pain and burden of calibration?

Element 4.11, Control of Inspection,
Measuring, and Test Equipment, causes
big trouble for many companies. It is
probably the single most dreaded and
feared requirement in QS-9000, for the
following reasons:

> **CAPSULE ANSWER**
>
> If a gage calibration prob-
> lem exists, it is vital to
> start solving it early. It will
> always take more time than
> expected.

- A few companies have *never* had a formal gage calibration or confirmation system. It's hard to believe this situation exists in an environment
  of companies working with Big 3 plants or their Tier 1 suppliers, but it
  does.
- Other companies—even fewer than above—*seem* to have a fully compliant gage calibration system, but actually do not. What they have is a
  sham, a "sticker and snicker" system that gained wink-wink tolerance
  among supplier quality "auditors." These companies had best not count
  on getting approved in a QS-9000 assessment.
- Other companies—a large number, based on personal observation—
  have a gage calibration and confirmation system but are not especially
  strict about enforcing it. The record keeping may not be complete.
  Many gages float around the floor—calibrated, uncalibrated, barely
  functional, whatever.
- Some companies have traditionally neither confirmed nor calibrated
  instruments that (in QS-9000) *must* be confirmed or calibrated. One
  large category is employee-owned instruments. Another is measuring
  instruments that are not often thought of as gages but qualify for calibration or for measuring conformity to requirements: tape measures,
  rulers, right angles, and so on.
- QS-9000 assessors *love* to audit gage calibration. The requirement acts
  like a magnet on many of them, probably because gage calibration is relatively easy to audit. It tends to be black and white; little interpretation
  is needed. An instrument is either calibrated, with appropriate records,
  or it is not.

When companies start to implement QS-9000, they find that this requirement produces a never-ending Maalox moment. How can the pain of compliance be reduced? Here are some commonsense ideas:

- If there is a known problem with this issue, do not duck it. Make it a priority item from the beginning of implementation. Assign a champion—a person with real clout—to handle it. Invest the necessary resources of time, people, and equipment in solving the problem. Make a plan and follow it, and *get started early!* The worst mistake a company can make is to close its eyes, clench its fists, and pray that the problem will just go away.
- Start at the beginning. Look at all the control plans and other manufacturing documents, and list the following:
  - Measurements being made to verify conformance to requirements.
  - The precision level required for each.
  - The best device for reaching that precision level.
  - The device actually being used (if different from the above).
- Scour the manufacturing floor for devices. Sweep through the place like Sherman through Georgia, and make a list of every single measuring device used for any conceivable purpose. Label or mark each one, to make it easy to spot any that have been missed. Record these data:
  - Name of device.
  - Serial number (if any) or part number (if any).
  - Ownership.
  - Area found/area of use.
  - Status: active (in use) or inactive (not used for current production).
  - Condition: calibrated, suspect, functional, nonfunctional, dysfunctional, broken.
- If a gage record system already exists, update it with the above information and flush out records of instruments that are no longer present.
- If no gage record system exists, start one. Use one of the many excellent PC-based gage control systems.
- With the help of quality and/or lead manufacturing personnel, make a study of the gage usage that has been found. Look for opportunities to *reduce the number of gages in use.* Are multiple gages being used in an area, to measure the same type of characteristic? Are employee-owned gages in use where company-owned gages are available? The more the population of gages is reduced, the easier (and cheaper) the ordering of a system will be.
- Rank the instruments according to the following priorities:
  - In use for current production.
  - Needed for expected production (e.g., inactive parts) but not in use now.
  - Not needed for any anticipated production.

- Take a hard look at the "not needed" category and simply discard as many of its members as possible. The balance, plus those in the "not in use now" category, must be marked "not calibrated" and locked up or otherwise made unavailable for use.
- Focus on devices in use for current production. Have the company's masters (blocks, surface plates, and so on) calibrated by an outside service (with all necessary certifications).
- Determine how to calibrate or confirm the rest of the devices. In a time crunch, they can be calibrated by outsiders—at a cost. Consider hiring temporaries until proper control is established.
- Write into the procedure that the "inactive" devices that have been locked away will be calibrated only as, and when, they are needed for current production. This entirely acceptable strategy will spread out the expense and work of calibrating them.
- Some devices cannot be calibrated (adjusted). They can only be confirmed as meeting a specified standard. Examples are: rulers, scales, and tape measures. These need to be inspected for damage and wear, and compared to some other defined standard, at reasonable intervals, to ensure their accuracy.
- Plan how to handle ongoing calibration needs, with internal resources or by utilizing outsiders. Update the procedure to reflect the method that will be used.
- Implement the plan.

64. What are the requirements for analyzing measurement and test equipment systems?

The Standard requires that statistical analyses be carried out on each measurement system listed in the Control Plan, to determine how much variation exists in the measurement data.

> **CAPSULE ANSWER**
>
> Measurement systems analyses ("gage R & R studies") study the variation in the measurement systems listed in a company's control plan.

Such studies are often called "gage R & R" studies because *repeatability* and *reproducibility* are two of the most common characteristics checked in these types of studies.

Whatever system is in place for checking the accuracy of individual instruments, it is commonly called a "gage calibration" system (Question 62). Measurement systems analyses go well beyond calibration. As a rule, they are carried out before new processes are initiated, as part of advanced quality planning (Question 30) and the Production Part Approval Process (PPAP; Question 39). They check the validity of *whole measurement systems* rather than *individual instruments*.

Why are such studies important? Because data are essential to controlling and improving the manufacturing process. And data are only as good, only as reliable, as the measurement system that produced them.

Assume a caliper is used to inspect a part or to gather readings on a critical characteristic in order to measure variation. The decisions made as a result are influenced by, first of all, the condition of that particular caliper:

- Has it been damaged?
- Is it worn out?
- When was it last checked?
- Does its accuracy vary from one part of its operating range to another?

The decisions are also influenced by the people using the caliper:

- Does a particular operator use it exactly the same way each time?
- Is there variation in the ways different operators use the instrument?

The decisions may also be influenced by the environment in which the device is used. Do the readings vary based on the temperature of the surrounding area? The humidity? The noise or dust level?

Measurement systems analyses are highly technical statistical studies designed to answer these and many other questions. They help to determine:

- Whether a particular measurement system and/or method is suitable for the desired purpose.
- The amount of variation that exists in the system.

This information helps with wise selection of measurement systems and accurate interpretation of their results.

Such studies are worth a book of their own: *Measurement Systems Analysis Manual*, published by AIAG and referenced by the QS-9000 Standard. Any company that does not presently do any form of measurement systems analysis should:

- Get the book noted above.
- Contract for the services of someone (preferably a staff member) who has extensive statistical knowledge and experience, to design and operate a gauge R & R system.

TECHNICAL REQUIREMENTS

1. Analyze statistically each measurement and test equipment system listed in the company's Control Plan, to identify its inherent variation.
2. Maintain appropriate evidence that this is being done.
3. Use methods and acceptance criteria that conform to the *Measurement Systems Analysis Manual*. Others may be used if approved by the customer.

TECHNICAL GUIDELINES

- Management must identify the statistical properties that are most important for the ultimate use of the data.
- Management must ensure that those properties are used as the basis for selecting a measurement system.
- Two phases or levels of testing exist:
  - Phase 1: Does the measurement system have the necessary statistical properties? What environmental factors affect the system?
  - Phase 2: Once in use, does the measurement system continue to have the necessary statistical properties?
- Measurement systems must be in statistical control (i.e., variation is from special causes, not from common causes).
- Measurement system variability must be small compared with:
  - Manufacturing process variability.
  - Specification limits.
- Measurement increments must be no greater than one-tenth of the smaller of either process variability or specification limits.
- Studies can check for five categories of measurement system error (the first three are the most common):
  - Bias (accuracy). A study of how accurate the measurement system is. The accuracy of the instruments in regular use is compared with measurements taken by "a higher level" of measurement equipment.

- Repeatability. A study of any variation in measurements taken with one instrument, used several times by one operator, to measure the same characteristic on the same part.
- Reproducibility. A study of any variation in the average of measurements made by different operators using the same instrument to measure the same characteristic on the same part.
- Stability. A study of the total variation in measurements taken with a measurement system on a single characteristic of the same master, or parts, over an extended period of time.
- Linearity. A study of the accuracy of an instrument over its entire operating range.

REQUIRED DOCUMENTS

1. Procedure for measurement systems analyses.
2. Possibly, work instructions for specific studies.

Improving the Process

65. What kind of controls over nonconforming product does the Standard require?

QS-9000 is a prevention-based system, so there should be no nonconforming product. But this is the real world, and, from time to time, a process will generate nonconforming or suspect product. When that happens, the task is to keep it from reaching the customer. Ideally, it should not proceed even one step further in the process.

CAPSULE ANSWER

A company is required to control nonconforming product and keep it from reaching the customer. This requires a system for handling and resolving all instances of nonconforming product.

The Standard requires a documented process for identifying and controlling nonconforming material. And the process applies to *suspect* material as well.

"Control" in this context means keeping such material from being accidentally mixed in with "conforming" material. The Standard recommends strongly that such material be segregated from the process flow in a foolproof manner. Disposal is done according to stated procedures. A company may, of course, scrap the material. But it may also regrade it (name it something else, so that it fits some other customer specification).

Most often, the material is reworked or repaired, to return it to customer spec. When this is done, the Standard requires:

- Documented procedures.
- Reinspection after rework or repair.

The customer may be asked to accept the material "as is." This alternative is permitted as long as documented customer approval is obtained and the Production Part Approval Process (PPAP; Question 39) applicable in such cases has been followed.

A documented and prioritized plan for reducing nonconforming product is required. The benefits of investing heavily in prevention methods are clarified by this requirement.

TECHNICAL REQUIREMENTS

1. A company must implement a documented procedure to prevent the following types of product from being used or installed:
   - Nonconforming.
   - Suspect.
   - Subcontractor product that is either nonconforming or suspect, as covered under Production Part Approval Process (PPAP; Question 39) rules.
2. The procedure must define the function that has the authority to review and dispose of nonconforming product.
3. The procedure must call out how such product is:
   - Identified.
   - Documented.
   - Evaluated.

- Segregated (when practical).
- Disposed of (see below).

4. Disposition may take the form of:
   - Scrapping.
   - Regrading for other applications.
   - Repairing or reworking, so that the product meets specified requirements. This must be done:
     - In accordance with documented instructions that are accessible to appropriate personnel.
     - With subsequent inspection, as spelled out in the quality (control) plan.
     - In a way that prevents rework from being visible on the exterior of a product supplied for service applications, except where approved by the customer.
   - Acceptance by the customer, either as-is or with repair. In this case:
     - The applicable Production Part Approval Process (PPAP; Question 39) must be followed.
     - The condition of the product as shipped must be recorded, as well as the customer's acceptance, the expiration date, and/or the quantity authorized.
     - When customer approval has expired, the original specifications (or a changed version, if approved) must be adhered to.
     - Nonconforming product shipped pending customer approval must be identified as such on the shipping label.

5. A system must be in place to quantify, prioritize, and reduce nonconforming product, and to track progress toward each of those goals.

TECHNICAL GUIDELINES

- Document the process for dealing with nonconforming product. Organize it to kick in immediately upon receiving indications that products or components do not, or may not, meet specified requirements.
- To prevent inadvertent use or installation of nonconforming product:
  - Promptly identify the units or batches involved.
  - As necessary, implement a process to reexamine previous lots.
  - Separate the defective product from the rest.

- Where the complexity of required storage conditions prohibits direct identification of conforming product, use documented or computer-based control systems for identification purposes (making certain to prevent inadvertent use or shipment of nonconforming product).
- Document the situation.
- Evaluate the nonconformity.
- Decide on and document its disposition.
- Ensure that the persons making this review are competent to evaluate the effects of their decisions on interchangeability, further processing, performance, reliability, safety, and aesthetics.
- Control movement of the nonconforming product until it is disposed of.

■ If the customer's decision is to accept, with or without repair or concession, document the reason and include appropriate authorizations and necessary precautions.

■ Notify all other affected functions.

■ Other outcomes are acceptable, but, ultimately, the safest and least risky course is to scrap any nonconforming product.

REQUIRED DOCUMENTS

1. Nonconforming product procedure.
2. Reduction procedure with associated plans (could be part of the above).
3. Nonconforming product records, especially those related to customer acceptance and PPAP issues.
4. Procedure or work instructions for:
   - Rework of nonconforming product.
   - Inspection of nonconforming product.

AUDIT ISSUES

■ Often, rework requires work instructions that differ from the instructions controlling the original work. Make sure the reworking instructions are available.

■ Segregation of nonconforming or suspect material is a very large audit issue. Do not depend on workers' knowledge or memory for the identification of such material:

– Have designated areas (containers, bins, etc.) on the shop floor and clearly mark them as receptacles for nonconforming material.
– Label, tag, or otherwise affirmatively indicate that a material is nonconforming or suspect.

■ "Hold cages" for nonconforming material are notorious for being repositories of material that is months or even years old. Make sure that such material is disposed of in a timely manner.

66. What corrective action systems must companies have in place?

> **CAPSULE ANSWER**
>
> Corrective action (and its twin, preventive action) are at the heart of the "continuous improvement" mechanism built into QS-9000.

QS-9000 requires a full-blown system for corrective action as well as preventive action (Question 67). It must be real, it must be effective, and it must be consistently operated.

When the system is applied conscientiously and consistently, with the involvement of everyone in the company, it results in major improvements throughout the process. This is fact, not fantasy. But it takes commitment, work, and a disciplined system.

The documented corrective action system must address customer complaints, product nonconformities, and nonconformities related to the process itself and the quality system (a very important point that is often overlooked).

The measures taken must be:

■ Appropriate to the magnitude of problems (the bigger the problem, the harder the company works on it).
■ Commensurate with the risks involved (i.e., when the cost versus return projection makes sense).

The procedure must call for:

■ Investigation and identification of causes.
■ Determination of corrective actions needed to eliminate the causes.
■ Controls (follow-up, at least) to make sure that the corrective action:
– Has actually been carried out.
– Is effective at eliminating the cause.

QS-9000 also requires a defined process for analyzing and testing returned products and for taking corrective actions on the causes. The company must also maintain records to show that the procedure is being consistently followed.

TECHNICAL REQUIREMENTS

1. A company must establish and maintain documented procedures to carry out actions that correct errors and eliminate causes.
2. Any such action must be:
   - Appropriate to the magnitude of problems.
   - Commensurate to the risks encountered.
3. When corrective action involves changes in procedures, a company must make those changes and ensure that they are effectively implemented.
4. A company must apply the following in its corrective action process:
   - Disciplined problem-solving methods (Question 17).
   - Testing and analysis of products returned from customer locations. Analysis must, where appropriate:
     - Lead to effective corrective actions and process changes.
     - Be recorded.
5. Corrective action includes:
   - Effectively handling customer complaints and reports of product nonconformities.
   - Investigating causes of nonconformities relating to:
     - Product.
     - Processes.
     - The quality system.
   - Recording results of the investigation.
   - Determining corrective actions needed to eliminate causes of nonconformities.
   - Applying controls to ensure execution and effectiveness of corrective actions.
6. A company must implement and record any process changes resulting from corrective actions.

TECHNICAL GUIDELINES

GENERAL

- Corrective actions can be directed toward in-process products, or products already delivered as satisfactory but then found to have nonconformities.
- Each member of the organization has a duty and responsibility to identify and report nonconforming products and services. Employees should make every effort to identify potential nonconformities before customers are affected.
- Management should empower everyone in the company with the authority to initiate corrective action requests. Management should refrain from attempting to limit the topics, but a screening process to weed out the clearly irrelevant requests is advisable.
- By procedure, responsibility for coordinating, recording, and monitoring corrective actions should be established. These responsibilities can transcend all facility functions.
- The results of corrective action are not technically required to be reviewed during management reviews. However, to do so is in the spirit of the requirements and is highly recommended.

ANALYSIS

- Analysis and execution of corrective actions may involve a variety of functions:
  - Sales.
  - Design and development.
  - Purchasing.
  - Manufacturing and related functions.
- Situations giving rise to corrective actions can include:
  - Product or service failures.
  - Faults in the process or its documentation.
  - Noncompliance with procedures.
  - Poor scheduling.
  - Process nonconformity reports.
  - Management reviews.
  - Market feedback.
  - Customer complaints.

- ■ Areas to analyze when formulating corrective actions can include:
  - – Inspection and test records.
  - – Nonconformity records.
  - – Process monitoring.
  - – Results of internal audits and/or management reviews.
- ■ When evaluating potential corrective action:
  - – Weigh the significance of the quality problem against its potential impact on processing costs, quality-related costs, performance, reliability, safety, and customer satisfaction.
  - – Analyze the problem and determine its root cause or causes. Statistical methods can be useful for this activity.
  - – Carefully analyze the problem from the standpoint of various process elements—design, quality plan, product and service specs of all related processes, and records.
  - – Observe a problem directly at the place where it was discovered.

EXECUTION

- ■ Corrective action may occur in two stages:
  - – Immediate positive action to satisfy the needs of a customer.
  - – Evaluation of the root cause of nonconformity to determine any long-term corrective action that is necessary to prevent recurrence.
- ■ Institute remedial action as soon as is practical, to limit the costs of reprocessing, reworking, reclassifying, or scrapping.
- ■ Make certain every corrective action is appropriate to the magnitude of the related problem.
- ■ Consider establishing a file or log that lists nonconformities and corrective actions, to help identify problems that have a common source.

FOLLOW-THROUGH

- ■ Monitor the effect of corrective actions to ensure the achievement of desired goals.
- ■ Document, as appropriate, permanent changes resulting from corrective action. Record needed changes in:
  - – Work instructions (including training).
  - – Manufacturing processes.
  - – Product specifications.

– The quality system.
– Procedures used to detect/eliminate potential problems.

REQUIRED DOCUMENTS

1. Corrective action procedure.
2. Corrective Action Request form.
3. Corrective action records (general).
4. Records of returned product analyses.

AUDIT ISSUES

■ Some companies limit corrective action to product nonconformities or customer complaints. These are important, but the Standard specifically requires that corrective action be applied to problems arising from *product, process, or the quality system.*

■ Corrective action must get at the root cause of problems. Simply dealing with the symptom is not sufficient to meet the requirements.

■ Follow-up is required. Assessors will track follow-up activities to satisfy themselves that the activities are real and effective.

■ Assessors will analyze corrective action for repetitions. If the same problem and/or cause comes up again and again, a judgment can be made that the corrective action has not been effectively implemented and the result may be noncompliance with the Standard.

67. What about preventive action?

QS-9000 is based on the principle of "prevention rather than detection." In that spirit, QS-9000 (and its parent, ISO 9000) requires defined preventive action.

This is a traditional puzzle point for many companies. How is preventive action separated from corrective action?

> **CAPSULE ANSWER**
>
> Preventive action is a documented process of identifying potential problems, eliminating their causes, and following up to confirm that the action is effective.

Preventive action is aimed at attacking and eliminating the causes of potential (rather than actual) problems involving the product, process, or quality system. Training and awareness are key here. Employees must be made aware of and kept alert to potential problems. They must be encouraged and rewarded for reporting those problems in the manner described in the company's procedure.

Preventive action can be an extension of corrective action. Where a corrective measure can be applied elsewhere in the process—to prevent a possible problem from occurring—it becomes preventive action and should be credited that way.

Preventive action, according to the QS-9000 requirements, is a distinct process of:

- Analyzing documented results in search of potential preventive actions.
- Identifying preventive actions.
- Implementing the identified actions.
- Following up to ensure that the actions are effective.

Many companies, saying "We don't have time," do not do any preventive action. Or, they do just enough of it to satisfy assessors (who often are not as rigorous as they should be on this issue). But, interestingly, companies always seem to have time to address and correct problems *after* they have occurred. They have plenty of resources for a "cure," but not enough for prevention.

Preventive action is one of the most proactive requirements of the Standard. Companies are sadly remiss if they do not fully avail themselves of it.

TECHNICAL REQUIREMENTS

1. A company must implement a documented procedure for preventive action.
2. The procedure must:
   - Use various sources of information to analyze and eliminate potential causes of nonconformities. These sources can include:
     - Processes and work operations.
     - Quality records.
     - Audit results.
     - Complaints.
   - Include steps for dealing with any problems requiring preventive actions.

- Specify how preventive actions are to be carried out.
- Provide for the application of controls to ensure that preventive actions are effective.
- Ensure that all preventive actions, especially those that result in procedure changes, are recorded and reviewed as part of the required management reviews (Question 31).

TECHNICAL GUIDELINES

- The Standard links preventive action closely with corrective action. The quality system should place emphasis on the identification of actual and potential quality problems and the initiation of remedial or preventive measures.
- The quality system should, according to the Standard, stress problem *prevention* "rather than dependence on detection after occurrence."
- Every person in the organization has a "duty and responsibility" to identify and report potential nonconformities "before customers are affected."

DESIGN FUNCTION

- One area the Standard singles out for preventive action activities is the *design function*. Effective design under a QS-9000 quality system is the ultimate in preventive action: in effect, it anticipates and solves problems before they occur. Indeed, the Standard says that preventing defects at the design stage "is less costly than correction during delivery."
- The Standard recommends that the design function carry out preventive action measures as follows:
  - Implement design reviews for each phase of design.
  - Validate that the delivery process meets customer requirements.
  - Plan for variation in demand.
  - Anticipate the effects of possible systemic/random failures, and develop contingency plans for them.
  - Conduct experiments to understand the technical condition of processes related to quality of products. Pay attention not only to removing deficiencies when actually found, but also to establishing future maintenance needs.

WITHIN THE PROCESS

- The Standard stresses preventive action within the process itself.
- The Standard requires:
  - Machine capability and instrument calibration (Question 62).
  - The establishment of a program of preventive maintenance (Question 57).
- The Standard recommends measures for early warning of out-of-control operating conditions in the production process, to prevent delivery of nonconforming products.

CLOSING THE LOOP

- When preventive actions are taken, it is important to confirm that the actions work as effectively as intended. The Standard recommends the following steps:
  - Make necessary changes to methods of design, development, manufacturing, packaging, and related activities.
  - Revise product specifications.
  - Revise the quality system.
  - Initiate preventive action to a degree that is appropriate to the magnitude of potential problems.
- The Standard recommends documentation changes. When a preventive action results in procedural changes, these should be documented in:
  - Work instructions.
  - Training materials.
  - Documents controlling manufacturing processes, production specifications, and/or the quality system.

REQUIRED DOCUMENTS

1. Preventive action procedure (could be part of Corrective Action procedure).
2. Preventive Action Report form (could be combined with Corrective Action Request).
3. Preventive action records.

AUDIT ISSUES

Although corrective and preventive action processes are linked, preventive action must be a distinctly separate process. It can be (and often is) an extension of corrective action, but must be called out as preventive action in the procedures and related forms and analyses, to demonstrate that the requirement is being met.

68. Are statistical techniques mandated by the Standard?

Yes. A company must not only have a system for identifying the need for statistical methods, but it must also employ statistical methods in several key areas.

On the product side, the Advanced Product Quality Planning system (Question 30) requires identification of statistical methods for the control of critical characteristics.

> **CAPSULE ANSWER**
>
> The Standard requires a company to utilize statistical methods to control special characteristics, and to analyze company performance at meeting defined customer quality concerns.

Other requirements for statistical methods are at the strategic rather than the operational level. The Standard requires a system for compiling data related to, and analyzing trends concerning, quality, operational performance, and so on, for key product and service features. This means a company must:

- Identify the product and service features that are of major importance to the customer.
- Analyze in statistical form how well the company is meeting the quality requirements related to those features.
- Prioritize customer-related problems so that meaningful solutions can be generated.
- Apply data analysis to business decisions and long-range planning.

The Standard also states that employees, as appropriate, should be knowledgeable in certain basic statistical concepts.

TECHNICAL REQUIREMENTS

1. The Standard requires a company to identify the need for statistical techniques to establish, control, and verify:
   - Process capability.
   - Product characteristics.

     This identification should be done during the Advanced Quality Planning process (Question 30).

2. The Standard also requires a company to establish and maintain (and to include in the Control Plan) documented procedures to implement and control the statistical techniques so identified.

3. Throughout the organization, employees, as appropriate, should demonstrate knowledge of basic statistical concepts, such as:
   - Capability.
   - Control.
   - Overadjustment.
   - Stability.
   - Variation.

4. For key product and service features, the company is required to:
   - Document trends in:
     - Current quality levels.
     - Operational performance.
     - Quality generally.
   - Compare or benchmark the trends against competitors, as appropriate, and/or against other benchmarks.
   - Compare trends with progress toward business goals.
   - Translate these analyses into information to support:
     - Development of priorities for prompt solution of customer-related problems.
     - Identification of key customer-related trends.

5. Ford suppliers are required to implement Ford's QOS (Quality Operating System), using standard tools and practices to manage the business. (See Ford's *QOS Assessment and Rating Procedure.*)

TECHNICAL GUIDELINES

GENERAL

- Choose metrics that best fit the situation; several different types of metrics can be used for several different situations.
- Statistical techniques, such as statistical process control (SPC), may be supplementary requirements of particular customer contracts.
- Consider training supervisory, technical, and nontechnical personnel in statistical techniques in these departments:
  - Marketing.
  - Procurement.
  - Process/product engineering.
- Encourage suppliers to apply statistical methods to control measurement processes.

TYPES OF TOOLS AND METHODS

- Records of the application of statistical tools may effectively demonstrate conformity to quality requirements. Types of statistical tools include:
  - Analysis of variance.
  - Cusum techniques.
  - Design of experiments/factorial analysis.
  - Graphical representations.
  - Quality control charts.
  - Regression analysis.
  - Statistical process control.
  - Statistical sampling.
  - Tests of significance.
  - Treatment of autocorrelated data.
- For more statistical tools, see ISO/TC 69 and ISO Standards Handbook 3, Statistical Methods.

AREAS OF APPLICABILITY

- Identify and use various quantitative measures in product quality.
- Report metric values on a regular basis.

- Establish specific improvement goals in terms of the metrics.
- Take remedial action if metric levels go outside target levels.
- Determine the quality of the development/delivery process:
  - Measure the development process in terms of milestones.
  - Measure the effectiveness of the development process at reducing the probability that faults are introduced and/or remain undetected.
- Establish and maintain a documented process for applying statistical methods in:
  - Capability studies.
  - Data analysis.
  - Defect analysis.
  - Forecasting.
  - Longevity and durability prediction.
  - Market analysis.
  - Problem analysis.
  - Process control/process capability studies.
  - Process improvement.
  - Product design.
  - Quality levels in sampling inspection plans.
  - Quality measurement as an aid to decision making.
  - Reliability specification.
  - Risk analysis.
  - Safety evaluation.
  - Statistical analysis of process/product performance.
  - Understanding customer needs.

REQUIRED DOCUMENTS

1. Procedure for identification of statistical techniques.
2. Procedure for analysis and use of company-level data.
3. Procedures or instructions (as needed) for application of statistical techniques.
4. Reports and analyses defining trends, and so on, in quality and operational performance (including baselines).

AUDIT ISSUES

- An audit of the Advanced Product Quality Planning system will cover identification of required statistical methods within the production process.
- Many QS-9000 candidates already have some form of performance measurement system. The existing system can easily be adapted to meet the requirement for analysis and use of company-level data.
- Mere data collection is not enough (and is one of the non-value-added pitfalls of this activity). A company should show how it is applying the information to address customer needs and improve its performance.

Post-Process Functions

69. Does the quality system cover handling, storage, and other post-process functions?

Yes. There are several very significant requirements covering handling, storage, packaging, and delivery processes. Overall, the intent of these requirements is twofold:

> **CAPSULE ANSWER**
>
> QS-9000 has strict requirements in this area, particularly for 100 percent on-time delivery to customers, documented inventory management systems, and shipment notification systems.

1. To ensure that product quality is preserved throughout the process and during the delivery sequence.
2. To ensure that the product reaches the customer when, and in the condition, required.

There is a great deal of flexibility in the handling requirements. The level of compliance depends strongly on the type of product being provided. In the area of storage, packaging, and delivery, the Standard gets pretty strict:

■ Storage of raw materials, subassemblies, and finished goods must be controlled. Implied here are:
  – Designated storage areas.
  – Procedures for checking product in and out.
  – Documented inventory management system.
■ Packaging must meet customer requirements (usually defined both in a general sense and on an order-by-order basis).
■ The company must establish a goal of 100 percent on-time delivery to customer requirements (Question 70).

For the most part, these requirements are fairly familiar to companies that have been Big 3 suppliers all along. If anything, QS-9000 puts some additional teeth into them.

TECHNICAL REQUIREMENTS

1. Implement a documented procedure for product handling, storage, packaging, preservation, and delivery.
2. Provide handling methods that prevent damage or deterioration.
3. To prevent product damage or deterioration, use designated storage areas or stockrooms pending use or delivery.
4. Establish a documented inventory management system to:
   ■ Ensure stock rotation.
   ■ Minimize inventory levels.
   ■ Optimize inventory turnover time.
5. Specify approved methods for authorizing receipt and shipping to and from delivery areas. Meet customer requirements for shipment, including:
   ■ Transportation mode.
   ■ Routings.
   ■ Containers.
6. To ensure conformance to requirements, control packaging and marking processes, including the materials used:
   ■ Follow the customer's packaging guidelines and standards.
   ■ Implement a system to ensure that materials are labeled in accordance with customer requirements.
7. Preserve and segregate product, via approved methods, while it is under the company's control.
8. Where specified by contract, protect product through its delivery to its final destination.

9. Chrysler suppliers must be familiar with Chrysler packaging and shipping instructions. Refer to the following Chrysler documents:
   - *Packaging and Shipping Instructions.*
   - *Shipping/Parts Identification Label Standards.*

TECHNICAL GUIDELINES

- The quality system should provide for proper planning, control, and documentation of the status and movements of all product: incoming, in-process, and finished.

HANDLING

- Consider provision of transportation units and maintenance of handling equipment.

STORAGE

- Consider:
  - Physical security and environmental conditions.
  - Periodic inspections for deterioration.
  - Incorporating expiration dates and stock rotation procedures.
- Specify appropriate methods for ensuring shelf life and avoiding deterioration.
- In compliance with requirements, check stored product at appropriate intervals.
- Provide for correct pallets, containers, and so on, to prevent damage from vibration, shock, or abrasion.
- Check processed materials in storage for deterioration, contamination, undesirable separation, or reaction.

PACKAGING

- Protect each product against damage and deterioration until the company's responsibility for the product ends.
- Meet all contract requirements for labeling.
- Markings and labels should be legible, durable, and in accordance with specifications.

- Identification should be intact from time of initial receipt to delivery at final destination.
- Markings should be adequate to identify particular products in the event that recall or special inspection becomes necessary.
- For a product delivered as continuous flow (e.g., through a pipeline), where no marking or labeling is possible, develop appropriate methods for identifying the product.

REQUIRED DOCUMENTS

1. Procedure for handling, storage, packaging, preservation, and delivery.
2. Work instructions for specific quality-related tasks described in the above procedure.
3. Documented inventory management system (can be computerized).
4. Customer packaging standards (as appropriate).

AUDIT ISSUES

- In many companies, control of raw materials, subassemblies, and finished goods inventory can be an issue. Even though many companies have moved toward just-in-time (JIT) systems (due to customer pressure), some inventory is almost always on hand and must be adequately controlled.
- The 100 percent on-time delivery requirement is a major issue with QS-9000. Assessors check to verify that there is, in fact, a documented process for:
  - Understanding what customer delivery requirements are.
  - Tracking lead-time requirements.
  - Tracking delivery performance.
  - Taking corrective action when delivery performance falls below 100 percent.

70. What kind of delivery requirements does the Standard specify?

CAPSULE ANSWER

QS-9000 requires a system that ensures 100 percent on-time adherence to customer shipping/delivery requirements.

Most of the clauses of QS-9000 are driven by customer needs and requirements. Those terms come up again and again. Nowhere is this more evident than in the "delivery" section of Element 4.15.

In short, the Standard requires that a company's system must:

- Support 100 percent on-time shipment to meet customer delivery requirements.
- Take corrective action (and notify the customer) when shipment(s) are not made on time.

That's 100 percent on-time delivery, not 99.9 percent—a tough requirement. What's even tougher is that the requirement is carefully worded to ensure that a company really has a top-to-bottom system for the promised 100 percent performance.

The company must:

- Compare the customer's due date with its own manufacturing process, and establish the production lead times needed in order to ship on a date that allows enough time to meet the customer's delivery requirement. This alone is something that many companies still do by the seat of their pants.
- Except where the customer grants an exemption, transmit an online advance shipment notification (ASN) to the customer. (Most companies that are required to have this already have it.)
- Track delivery performance versus customer delivery requirements. This helps to:
  - Verify that the lead times are realistic.
  - Assess how well customer delivery requirements are being met.

The customer must be notified when delivery requirements cannot be met, and documented corrective action must be taken to improve the process so that 100 percent on-time delivery can be achieved.

The Big 3 are dead serious about this issue. It is a major audit target. A company would do well to invest serious resources in establishing systems that ensure consistent fulfillment of the requirement.

TECHNICAL REQUIREMENTS

1. The company must establish systems to support 100 percent on-time shipments that meet customer delivery requirements.
2. The company must systematically:
   - Identify the lead times needed to meet customer delivery requirements.
   - Evaluate and monitor adherence to lead-time requirements.
3. Except where waived by the customer, the company must transmit advance shipment notifications (ASN) via computer, at the time of shipment. (Objective evidence of a customer waiver must be provided.)
   - ASNs must match shipping documents and labels.
   - Should the online system fail, the company must have a backup method.
4. The company must meet customer requirements for mode of shipment, using customer-specified containers, mode of transportation, and routings.
5. The company must systematically track delivery performance versus customer requirements.
6. The company must communicate problems with delivery to the customer, as needed.
7. The company must implement corrective action to improve delivery performance, when 100 percent on-time shipment is not achieved.

TECHNICAL GUIDELINES

- Take into account:
  - Delivery time requirements.
  - Environmental conditions that may be encountered.
- Verify correctness and completeness of delivered items.
- Provide appropriate protection of the quality of the product during delivery, especially:
  - Product with limited shelf life.
  - Product requiring special protection.
- Ensure that deteriorated product is not shipped and put into use.
- Identify products that have limited shelf life or require special protection during transport or storage.

- Where processed materials are hazardous, consider health education and safety issues.
- Identify special hazards inherent in products delivered in continuous flow.

REQUIRED DOCUMENTS

1. Shipping procedure.
2. Work instructions, as needed, for packaging and loading.

AUDIT ISSUES

- A workable and documented system must be in place for estimating the production lead times needed to meet customers' delivery requirements. Support this system with analysis that tracks actual delivery performance versus customer delivery requirements.
- Be sure to take documented corrective action when shipping deadlines and/or customer due dates are not met.

PART FIVE

Implementing a QS-9000 System

71. What is the one key thing needed to implement QS-9000 effectively?

CAPSULE ANSWER

To get the job done, senior management must lead.

Management commitment, especially the consistent and persistent commitment of senior management. Without it, a company cannot succeed. With it, the company cannot fail.

Too often, however, senior management is not in the picture at all. Or, senior management starts the process, issues edicts, and then sits back to watch. Some executives seem to want implementation to happen by magic—"without changing our business." Some want it to happen for free.

Effective QS-9000 implementation cannot happen by magic, or without changing the business. And it most certainly cannot happen for free.

It happens because of a lot of hard work. There's no way around that. And it changes the business, ultimately for the better, because the principles behind the requirements are immutable. The requirements themselves are subject to interpretation, and not every requirement applies to every company. But the principles themselves are immutable. Adopting them changes the company eventually, for the better.

Effective implementation comes at a cost. The exact amount varies from company to company (Question 8). At some point down the road, the QS-9000 system will migrate from a cost to a benefit. But that does not happen until the system is approaching steady state and has started to become transparent: "The way we do things around here" instead of "This pain-in-the-rear QS-9000 deal."

Before that transparency happens, a company has to implement. *As long as a company remains in an implementation mode, QS-9000 is only a cost, a burden.* The company's goal, and collective job, is to implement as effectively and as quickly as possible. And, to repeat, the key is consistent, persistent commitment from senior management.

What are the elements of senior management's role? They are best described by addressing top executives directly, as follows:

- Get out in front from the very start. Declare your personal commitment. Acknowledge that this is something the company has to do to meet your customers' needs. This is not a negative move unless you make it one. Stress that implementation is a chance to improve the business.

- Appoint a key person as management representative (MR). Choose someone you trust, preferably an old-timer who knows his or her way around the company and the industry. The MR should be someone who is visibly important within the company, to send a powerful message: "We are really serious about this."

- Start with a plan. Implementation will never get done otherwise. Insist on a detailed plan, with schedules, goals, and milestones. Until your plan is complete, you are not ready to implement. And if you think you can implement "in our spare time," you are dreaming.

- Start with a budget. You can probably get a budget prepared free, if you're willing to shut down your company for a few months. But because few companies can afford the luxury of a temporary closing, the budget is going to cost you money. Get a detailed budget from a professional planner.

- Provide resources. Implementation is going to require some "extras," in the form of time and people, that you will not need after the system is implemented and you've reached steady state. Some needs will not be anticipated. Naturally, continue to require that such investments must be justified, but expect the expenses and account for them.

- Brace yourself. You are going to learn things about your company that you never knew. (So will every other employee, for that matter.) Some of these discoveries you are not going to like. How you react, how you deal with the unexpected, will be watched carefully by the people around you. Remain equable. View these situations as opportunities to improve. Address the problems and solve them in a disciplined way.

- Enforce the schedule. If the plan is realistic, the resources are on hand, and no hideous surprises emerge, there is no reason why your goal cannot be met on time. Review progress regularly. Question slippage. Enforce the schedule. Make your people deliver. Keep you eyes on the prize. Don't forget: the QS-9000 does not even begin to work for you until after implementation is done. Make implementation a phase, not a way of life.

- Expose yourself to every part of the system at some point along the way. Get familiar with the training. Look over the documents that are being written. Talk to employees at all levels and functions, over time and on a casual basis, to get their views on how the system is working. Your consistent expression of interest will do more to motivate them than all the edicts in the world.

- Learn what you must know, and do it sooner rather than later. Your long-term homework assignments are:
  - Gain an overall awareness of what the system is and how it works. The nuts and bolts can be delegated.
  - Learn the five Key Facts (Question 78), the quality policy statement (Question 81), and the processes for management review (Question 31), a business plan (Question 24), and analysis and use of company-level data (Question 23).

- Stay involved! Don't just announce the program, assign tasks, issue edicts, growl "Make it so," and retreat into the woodwork. Bring up the program at every opportunity: staff meetings, customer gatherings, trade shows. Make it part of your job.

72. When a company has more than one location, what's the best way to set up a QS-9000 system?

> **CAPSULE ANSWER**
>
> The way a quality system is structured in a multiple-location company depends in large part on the activities carried out at each location.

There are a couple of possible ways. Each company has to decide what will best satisfy its needs.

Begin by establishing what is actually done at each location. The Standard defines two types of locations:

1. A "site" is a location that has a production process and makes things. If it makes production materials, production or service parts, and/or

certain finishing services, it must be assessed and registered to QS-9000:

■ If it supplies directly to the Big 3 (Question 5).

■ If its customers are requiring it to register.

2. A "remote location" is one that does not produce things. Usually, remote locations carry out activities like these:

■ Accounting.

■ Administration.

■ Engineering.

■ Purchasing.

■ Sales.

■ Warehousing.

A remote location cannot be registered to QS-9000 independently. But each remote location that supports a site must:

■ Undergo registration assessment.

■ Be included in that site's QS-9000 registration.

Certain types of remote locations (not all of them) may be audited on a "sampling" basis during registration and/or surveillance assessments. The number of locations sampled, and the amount of time devoted to each, is at the discretion of the registration body. The rule of thumb is that such locations can be sampled as long as they do not add value to the dimensions or attributes of the product or service that the company provides. Sales or distribution locations are obvious candidates.

What are the options for a multiple-address organization?

1. Register each site (with the remote locations that support it) separately. This must be done if:

■ The sites have separate and distinguishable quality systems. If each has its own quality manual (Question 35), then the sites have separate quality systems.

■ Each site ships separately, and independently, to customers, and never (or virtually never) ships parts or components to another site. This implies totally separate processes, virtually mandating separate quality systems.

2. Obtain a "corporate scheme" registration—all installations (sites and their remote locations) are assessed and then registered under a single certificate. This is permitted if:

■ All sites work to a single quality system (i.e., a common quality manual) that is centrally managed. This implies a single quality manual and a management representative who oversees the quality system at all locations. Lower-tier documentation may vary from site to site, and local "quality representatives" might run things on a day-to-day basis. But the overall system must cover all locations. This approach is optimal when the sites and locations are within a reasonable geographical distance of one another.

If eligible, a company may best serve its long-term business interest by obtaining a "corporate scheme" registration. Here is why:

1. This type of registration could end up costing less money. The audit days required per site (Question 9) may be less than for individual site registration. The company would pay less for the registration audit and for surveillance audits. Only a careful evaluation of quotes from several registrars can confirm this.
2. If the sites interact a great deal, placing them all under a common, centrally managed quality system could improve communication, quality, and results. On the other hand, if they currently operate autonomously from a quality/management standpoint, placing them under a common quality system could be a wrenching experience and could add to the stress of doing the implementation. Under these circumstances, an exceptionally strong and determined senior management team would be needed to make the project fly.

As a practical matter, it would be wise to confer with at least one registration body on the pros and cons of a corporate scheme before proceeding with implementation. The registrar can look at the company's structure and system, review the available options, and provide cost figures.

73. QS-9000 sounds like a big nut to crack. Could a company go for ISO 9000 first, and then "upgrade" to QS-9000?

CAPSULE ANSWER

For companies with no deadline pressure, and with little if any quality systems or practices in place now, phasing ISO 9001/2 in first and then migrating to QS-9000 may make sense.

This approach has been discussed by many companies, and used successfully by a few. At first blush, it seems an easier route, especially when the two Standards are compared (Question 2). Alongside QS-9000, ISO 9000 looks almost like a walk in the park.

That notion is deceptive, however. For most companies, neither ISO 9001 (for design-responsible locations) nor ISO 9002 (for non-design-responsible locations) is easy to implement. ISO 9000 is much less prescriptive, does not mandate quite so many specific systems and practices, and is shorter. But it just looks easier!

For companies that have no documented, formalized quality system or defined practices in place and have to start from square one, ISO 9001/2 is tough—and QS-9000 is an even harder road. For these companies, a phased-in approach may make sense.

To decide whether to implement ISO 9000 first and then migrate to QS-9000, a company needs to answer a few more questions.

1. *How soon is the company required to be registered to QS-9000?*

A supplier of any of the following items directly to Chrysler or GM must be registered by July 1997 (Chrysler) or December 1997 (GM):

■ Production materials.
■ Production or service parts.
■ Finishing services.

If the company itself does not supply these items directly to the Big 3 but one of its major customers does, that customer may require the company to register to QS-9000, and may set a deadline for registration (if it has not set one already).

If a company's deadline for registration to QS-9000 (regardless of the source) is less than a year from now, it had better implement QS-9000—the whole thing, all at once. Odds are that there will not be time to do a phased implementation. However, if no customer is requiring registration to QS-9000 by a certain date, the company has much more flexibility.

2. *Does the quality system include most if not all of the usual automotive-related quality practices?*

Perhaps the company is already rated under Pentastar (Chrysler), Q1 (Ford), or Targets for Excellence (GM). If so, many of the most critical specific quality practices are in place in some form. Examples of these practices include:

- Production Part Approval Process (PPAP).
- Measurement system analysis, sometimes called gage repeatability and reproducibility (gage R & R) studies.
- Advanced quality planning (AQP).
- Problem-solving systems such as the Seven Disciplines (7D) (Chrysler).

A company that already has these processes has a significant head start on QS-9000. All other things being equal, it should probably go for QS-9000 without phasing in with an ISO 9001/2 system first. If none of these systems is in place today, it should consider implementing ISO 9001/2 first, and migrating to QS-9000 as a second step.

3. *How ambitious is the company?*

If it is not under deadline pressure and has none of the automotive quality systems described above in place today, the company should consider a phased approach to QS-9000. The phases would be:

- Implement ISO 9001 (or ISO 9002, if no sites are design-responsible).
- Obtain registration. The registration body must be informed of the plan to upgrade to QS-9000 as a second step. It will then ensure that the audit team that does the ISO 9000 registration audit is fully qualified under QS-9000 rules.
- Add the automotive-related requirements from QS-9000.
- Upgrade the registration to QS-9000. (This can be done during a scheduled surveillance assessment.)

A more aggressive and ambitious company may choose to go straight for QS-9000, all at once, without phasing in ISO 9001/2 first, even though it is under no deadline pressure and has no automotive quality practices currently in place. If that is the decision, go for it!

74. What is meant by "effective implementation" of the quality system?

QS-9000 requires that the quality system (and its documented procedures) be effectively implemented (Element 4.4.2b). But QS-9000 does not spell out what "implementation" is! Surprisingly, the ISO 9000 requirements and guidance documents do not say much about implementation, either. All they say is:

> The implementation of a quality system. . . . is most effective when those in the organization understand its intention and how it functions, in particular in the area of their responsibility and its interface with other parts of the system. [ISO 9000-2: 1993, Section 4.2]

According to the textbook (as well as most registration assessors), effective implementation is a rather short-range proposition. A quality system is deemed effectively implemented when employees:

■ Understand the company's quality policy (Question 22) as it relates to their own job functions.
■ Know where to find the procedures (and related documents) that affect how they do their jobs. These include:
 – Instructions and procedures that tell them how to do their individual day-to-day tasks.
 – Nonconforming product procedures.
 – Corrective/preventive action procedures.
 – Quality records procedures (as relevant).
 – Internal quality audit procedures (as relevant).
■ Know how to effect changes to these documents, as needed.
■ Work in a manner that is consistent with what the documents say.

This description suggests two things:

1. Many previous quality programs were never really "effectively implemented." For that reason, many companies have whole bookcases groaning with three-ring binders jammed with "procedures," "policies," and other documents that some employees know about, few understand, and hardly any follow. These "systems" were paperwork

exercises. Auditors—usually from customers—never really audited for effective implementation. They audited only to see that the required bits of paper were in place.

2. Writing a QS-9000 quality manual and procedures is only the end of the beginning, not the beginning of the end. Hard as "the writing part" is to get done, it is not the end! Not by a long shot.

To implement effectively, employees must be educated in the system. When everyone whose work affects quality has the knowledge set forth above, *only then* can the company begin to say that the system is effectively implemented.

With that level of effort, the company fulfills the textbook definition. If all it wants is to "get registered" to shake its customers off its back—and it is content to let the QS-9000 system be a cost, rather than a benefit—then it can settle for just "getting the certificate." But maybe it wants to do more. Maybe it is not content to let the QS-9000 "program" be just another expense item on its financial statements. Maybe it would like to leverage that expense into an investment that brings the company a return.

With that approach, effective implementation takes on a whole new meaning. A QS-9000 system that is *truly* effectively implemented is a dynamic, living, breathing system that:

- Fits a business like a glove.
- Adapts to change.
- Contributes consistently to improvement.

To get to that level, everyone (starting with senior management) has to work the system consciously, consistently, and persistently. Not just on registration audit day or management review day or surveillance assessment day, but *every* day. When everyone does that for two or three years, the QS-9000 system will become:

- Transparent within the company.
- Second nature to employees.
- Integral to the company's management.
- Key to strategic as well as tactical planning.
- A benefit rather than an expense.

These are the characteristics of a bona fide effectively implemented system.

75. What is the role of the management representative?

CAPSULE ANSWER

The management representative (MR) runs the key elements of the quality system, reports on its status to senior management, and usually handles liaison with the registration body.

The Standard sets forth fairly simple and specific requirements about the management representative (MR). The MR must be:

- A member of company management.
- Appointed by "the management with executive responsibility" (president, general manager, or the equivalent).

Is a company required to appoint its quality manager as the management representative? Often this is the most practical idea, but it is not mandatory. The MR should be the most appropriate person for the job. For example, if the quality manager is new to the company, or is close to retirement, or lacks confidence, clout, and competence, it would be wise to choose someone else.

In terms of responsibilities, the Standard simply says that the MR must have the "defined authority" to establish, implement, and maintain the quality system. This is a lot of authority! For this reason, it is tough to justify—and even tougher to sell, conceptually, to a registration assessor—a management representative who is two or three reporting levels away from the president or the general manager.

The Standard also makes the MR responsible for "reporting on the performance of the quality system." Although the link here isn't specified, the reference is to the management reviews (Question 31) required by the Standard. Senior management does the actual reviewing, but the MR is the person doing the reporting. He or she generally facilitates the management review meetings, documents them, and takes action on the results.

The Standard suggests that the MR could have responsibility for acting as liaison with "external bodies" on matters involving the quality system. This usually means handling the relationship with the registrar (Question 21). What the Standard is merely "suggesting" is something that is almost universally done. Most MRs handle all dealings with the registrar: scheduling, reporting, corrective action, and so on.

In the real world, the MR has these other specific responsibilities:

1. Running the corrective/preventive action system. The paperwork can be delegated, but the MR usually:
   - Screens corrective action requests.
   - Assigns them to responsible persons for action.
   - Enforces deadlines.
   - Follows up on results.
   - Analyzes the activity.
   - Reports to management, as required.
2. Running the internal quality audit system (Question 32). The MR does not do any actual auditing (the independence clause forbids this). But the MR takes care of these related activities:
   - Plans and schedules audits.
   - Provides auditors with needed documents and special instructions.
   - Evaluates audit reports and corrective action requests.
   - Mediates conflicts over audit results.
   - Follows up on corrective actions, as needed.
   - Schedules follow-up audits, as needed.
   - Analyzes the activity.
   - Reports to management, as required.

In some situations, the MR gets additional duties. One common syndrome occurs when a company does not have one or more systems or processes required by QS-9000. For example, if a company does not have a consistent and disciplined record-keeping system, and must implement one, the MR may get the ongoing duty of managing it. This add-on should be avoided. The MR should focus his or her time and attention on, first, implementing the system, and second, managing it via the systems described above, without becoming a dumping ground for duties that other people do not want.

The MR's role is vital and covers much ground. Yet, a full-time MR is a rare bird indeed. Most MRs accomplish their duties comfortably in about 25 percent of their worklife time.

TECHNICAL REQUIREMENTS

1. The management representative (MR) is a member of senior management, appointed by someone on the highest level of executive responsibility.

2. The MR must have defined authority to:
   - Ensure that a quality system, complying with the Standard, is established, implemented, and maintained.
   - Report to company management on the performance of the quality system.
   - Aid in the improvement of the quality system.
3. The MR may also have the responsibility for acting as liaison with external parties on matters pertaining to the quality system.

REQUIRED DOCUMENTS

The identity of the management representative must be defined. This is usually done in the *Quality Manual*.

AUDIT ISSUES

Assessors look askance at management representatives who do not have direct access to, and/or the confidence of, the senior manager of the company. To accomplish the objectives mandated by the Standard, the MR must have a considerable amount of authority and autonomy in the organization.

76. What elements of QS-9000 are most often *not* found in existing companies?

This true fact may be hard to believe: The typical company *already meets* the majority of QS-9000 requirements. The overall problem is twofold:

CAPSULE ANSWER
Most companies already meet most requirements, but several requirements are almost never met at the outset. The number and range of these depends, in large part, on whether the company has been an automotive supplier before.

1. The company's systems are usually not structured or documented as required by the QS-9000 Standard.
2. The system is often not thoroughly implemented or consistently followed.

As an example, the Standard requires that "contracts" (customer orders) be reviewed to ensure that:

- All required information (customer requirements) is present and accurate.
- The company has the ability to meet the customer requirements.

Every company already does this; it has to. So, technically, every company is in compliance on contracts. But there are some gaps:

- The means by which the company is in compliance is, typically, not formalized or documented.
- The required amount of record keeping is often not present.
- Different employees are doing the tasks in slightly different ways.

In large part, then, the process of implementing a QS-9000 system is a matter of formalizing and documenting what is already being done. This has the natural effect of making things more consistent, because everyone whose work is affected by a procedure is required to work in accordance with the procedure.

In the real world, the typical company, at the outset, lacks compliance to several key elements. The number and identity of these elements depend on whether the company has been an audited supplier to the Big 3.

Here are the requirements that are almost never met by a typical company making early attempts toward QS-9000 registration:

- *Control of customer-supplied product* (Question 41). Items supplied by customers must be protected and accounted for, and problems with such items must be reported to the customer.
- *Corrective and preventive action* (Question 66). A process must be in place to: identify the causes of existing and potential quality-related problems, eliminate the causes, and verify the effectiveness of the actions.
- *Document and data control* (Question 38). Documents affecting the quality system must be approved and controlled so that the latest approved issues are available to those who need them.
- *Internal quality audits* (Question 32). The entire quality system must be audited on a defined and documented basis, by trained independent auditors, for compliance, suitability, and effectiveness.

- *Management review* (Question 31). Management is required to review the status and effectiveness of the quality system at defined intervals.
- *Quality planning* (Question 30). The means by which customers' quality requirements are met must be planned and documented.
- *Statistical techniques* (Question 68). The means by which appropriate statistical methods are selected and used must be defined and documented.
- *Subcontractor development* (Question 43). The company must perform subcontractor quality system development using QS-9000 as the basic requirement.

Those "gaps" are usually present, regardless of whether the company has been a supplier to the Big 3. If the company has *not* previously been an audited supplier to the Big 3 (or its direct suppliers), it most likely lacks several additional systems:

- *Advanced product quality planning* (Question 30). The company must have a comprehensive system that leads to the development of Control Plans (Question 55) and includes the following subelements:
  - Failure Mode and Effects Analysis (FMEA)—design FMEAs (if design-responsible) and process FMEAs.
  - Identification and control of special characteristics (Question 49).
  - Production Part Approval Process (PPAP) (Question 39).
- *Continuous improvement* (Question 27). The company must have a comprehensive philosophy, backed by a defined process, for identifying and following through on continuous improvement projects.
- *Customer satisfaction system* (Question 41). The company must have a process for determining what customer satisfaction is, tracking it, and benchmarking it.
- *Delivery performance system* (Question 70). The company must have a process to ensure 100 percent on-time delivery to meet customer requirements.
- *Measurement systems analysis* (Question 64). This function identifies the amount of variation inherent in measurement systems used to monitor quality characteristics.
- *Problem-solving methods* (Question 17). These methods are to be used, in customer-prescribed format, to deal with internal or external nonconformances.
- *Process capability studies* (Question 58). Preliminary and ongoing studies are conducted to assess the ability of the process to meet quality requirements.

- *Process monitoring and operator instructions* (Question 37). These are required for all process activities.
- *Quality records retention requirements* (Question 33). Retention intervals are specified in the Standard and also by individual customers.

77. How should a company plan and schedule an implementation program?

> **CAPSULE ANSWER**
>
> There are four generic phases to QS-9000 implementation, and the steps carried out have a logical order.

Every implementation is different. Even within industries, companies differ widely, and implementing QS-9000 is, to an extent, a custom-designed effort.

Even so, each implementation proceeds through a set of fairly standard phases. There is a logical order to the process. Some parts of it are intuitively obvious. Others must be learned through experience.

These phases are not necessarily discrete, as shown in the sample spreadsheet in Appendix B. They can overlap quite a bit. The phases will not necessarily fit the exact timetables estimated for them, either. A certain amount of "winging it" is to be expected here.

PHASE I: PREPARATION

MANAGEMENT COMMITMENT

Why does this come first? Because without it, a company might as well not invest another nickel's worth of time or resources on the process. Senior management must commit itself to the process at the outset, and must stay involved as the process moves forward.

PLANNING, SCHEDULING, STAFFING, BUDGETING

A schedule becomes a management tool by using a spreadsheet. At least 12 columns should be allotted, one for each month. Most companies work backward from the date by which they must be registered. In some cases, the date is prescribed by customers. For example:

- Chrysler requires registration by July 1997.
- GM requires registration by December 1997.

Here are the steps for creating a spreadsheet schedule:

1. Find out whether other customers have deadlines for registration. If no customer is imposing a deadline at this time (lucky you!), set target dates and performance intervals that will keep the process moving forward.

2. Write the target month for registration in the last (twelfth) column of the spreadsheet. Then count back a minimum of 4 months (or, preferably, 6 months) from the target month. Subtitle that month "D-Day." It is the month by which the QS-9000 system must be designed, in place, and operating. It does not have to be fully functional, debugged, or "perfect." (No system is ever "perfect," not even years down the road.) But it has to be in place and running a minimum of 3 months (preferably, 6 months) before the registration audit. This lead time offers sufficient evidence and records to prove that the system is in place.

3. Back up 6 months from the D-Day column. On average, the implementation process must start in this month, in order to have enough time to get the job done. About one year is the average time needed to implement the system, from start to finish.

4. If the month that emerges as the "start date" is already two or three months in the past, do one or two things:
   - Compress the schedule to meet the required deadline.
   - Negotiate (if necessary) to extend the deadline.

Next, the project needs to be staffed. Senior management begins by appointing a management representative (MR; Question 75) to lead the effort. The MR should have a deputy, as well as dedicated staff support. Other functions need to be drawn into the process at this point, including choices for a Quality Steering Team (QST; Question 79).

Budgeting can get underway at this point. Begin by plugging in numbers for the following:

- Overview training. Thirty minutes of time for every employee in the organization.
- Orientation training. Ninety (?) minutes of time for every employee in the organization. (This is a hard one to call.)

- Documentation writing training. Two days of time for perhaps 12 to 15 key operations people from a cross-section of the organization.
- Internal quality audit training. Two days of time for the number of employees that represents about 10 percent of the total head count.
- Internal quality audit costs (very iffy). An average of 4 hours each for 2 auditors (total of 8 hours per audit), times the number of standard operating procedures in the system (at least 20; could be as many as 26). At least one complete cycle of internal quality audits must be completed before the registration audit.
- Equipment and supplies. Include a good computer with word processing software, and the services of someone who knows how to use it.
- Registrar services (Question 8).

(Question 8 also has more information on implementation costs.)

PROJECT LAUNCH

A launch begins by getting QS-9000 orientation training for senior management—the CEO, the general manager, or whoever is in charge of the organization—and for the Quality Steering Team (QST). Normally, this training can be completed in a day, or less.

Next, senior management needs to create the company's Quality Policy Statement (Question 81). This may take some time. It is subject to change, but senior management should put sincere effort into making the initial statement:

- Relevant to the needs and expectations of customers.
- Specific to the company.

Senior management then announces to the organization the process, the plan, and the schedule. At this point, the response of the employees will most likely to be a collective "HUH?" Their understanding will develop in subsequent phases.

PHASE 2: DOCUMENTATION

The process for creating the quality system documents is discussed in Questions 82 and 83. These are the elements that must be created:

- The initial Standard Operating Procedures (SOPs). These are needed to facilitate the writing and issuance of the other documents in the system.
- The *Quality Manual* (Question 35), which must then be approved and issued.
- The balance of the Standard Operating Procedures (SOPs).
- The necessary work instructions.

PHASE 3: TRAINING

Two types of training are involved here. The first type is training in the functions needed to implement and run the QS-9000 system. These functions include:

- Documentation writing.
- Internal quality auditing.

The other type of training is awareness and orientation training. It is very important that this is done as a planned, phased activity. The worst mistake a company can make is to hurl the QS-9000 system at its people at the last minute!

Question 78 has details on training.

PHASE 4: CRUNCH TIME

This is the period between D-Day—when all elements of the system have been set up and are functioning—and registration audit time. Think of it as a "shakedown cruise" for the QS-9000 system—or as white-knuckles time. There won't seem to be enough time to get everything squared away and running properly.

During crunch time, all employees should be "working the system." More formally, they should be:

- Aware of the procedures that affect how they do their jobs.
- Working in a way that is consistent with those procedures.
- Requesting changes to documents where they feel changes are needed.
- Reporting quality-related problems via the corrective and preventive action system.
- Dealing with and resolving quality-related problems via the corrective and preventive action system.

Other ongoing activities during Crunch Time include:

- *Management reviews.* After a system is established, two or three of these reviews are conducted each year. During implementation, a formal management review should be held at least once a month, to keep senior management tightly focused on the progress of the implementation. Frequent reviews also prevent surprises.
- *Orientation training* (Question 78). This activity should be wrapped up in the early part of this period.
- *Internal quality audits.* The audits tend to drive the process; they are the most important activity during Crunch Time because:
  - They get people focused on what the documents say.
  - They give people feedback on how they are doing.
  - They provide to senior management essential information on how the process is going.
  - They give employees some real-world experience in being audited—a valuable background when registration audit time comes.
- *Selection and scheduling of a registration body.* This process should begin prior to D-Day and be concluded shortly thereafter (Question 91).
- *Preassessment (Readiness Review).* This gives a company a final shot at fine-tuning its system before the registration audit (Question 92).
- *Registration assessment.* The big day!

78. What types of training should be provided to employees during the implementation process?

Training is a big part of successful implementation. Even if employees have been involved in automotive quality systems before, many aspects of QS-9000 will be new to them.

Two types of training are discussed here:

> **CAPSULE ANSWER**
>
> People must be trained to write quality system documents, and to be quality system auditors. But every employee needs awareness and orientation training on a phased, easy-to-understand basis.

1. Training that is specific to implementation tasks, which include:
   - Documentation writing training.
   - Internal quality audit training.
2. QS-9000 awareness and orientation training, which is where many companies drop the ball. This type of training must be provided on a

"phased" basis—a little at a time, over a period of time, in reasonable, easy-to-digest chunks. The worst possible approach is to call everyone into a room, dump a pile of manuals in their laps, order them to "learn all of this by Friday; we're getting audited," and stalk out of the room.

AWARENESS TRAINING

A very rudimentary "overview" session can be conducted with very large groups. The session should be:

■ Carried out within days of the project launch.
■ Thirty minutes (or less) in length.
■ Motivating as well as educational.
■ Reassuring (the company's whole world is not going to change).

The session should give all employees a snapshot of:

■ What QS-9000 is.
■ Why the company is getting involved.
■ How the system will affect them.

ORIENTATION TRAINING

Employees learn the key facts about the QS-9000 system that *every employee must know*. The critical issues on this training are:

■ It should be done about a month after the awareness training, but no more than 90 days afterward.
■ It should be conducted by the employees' direct supervisors, not by human resource or training personnel. The supervisors have to become "QS-9000 literate" in order to do the training.
■ It should take no more than one hour to complete.

The training should address the "Five Key Facts Everyone Must Know":

1. The name of the quality system (QS-9000).
2. Knowledge of the quality policy statement.
3. Location of quality system documents that affect employees.

4. The system for requesting changes to quality system documents.
5. The system for reporting quality-related problems for action (corrective and preventive action system).

Upcoming activities should be announced and explained:

- Internal quality audits.
- Registration audits.
- Surveillance audits.

Here is the kicker. Prepare a brief quiz consisting of maybe 10 or 12 questions on the above facts. Have the employees complete the quiz on an "open book" basis. They can even work on it in pairs, if they want to. Go over the answers and then have them turn the quiz in.

DOCUMENTATION WRITING TRAINING

This training is optional but helpful. The team selected to take the lead in writing the documents should have some rather intensive training in the following:

- The QS-9000 requirements.
- The documentation structure.
- The documentation writing/editing process.
- The company's approved document format.
- Rules for creating effective quality system documents.
- The theory and practice of document control.

This training can be completed in a two-day session.

INTERNAL QUALITY AUDIT TRAINING (QUESTION 32)

This is mandatory training for the team selected as internal quality auditors. Scheduling is critical. The training needs to take place right around the time the first Standard Operating Procedures (SOPs) are ready to be implemented. This continuity allows the company to set the audit schedule and start auditing promptly upon completion of the training, thereby maximizing its effectiveness.

Normally, the course runs for two days and addresses the following:

- QS-9000 requirements.
- Purpose of auditing.
- Strategic importance of auditing.
- The phases of a typical audit.
- Audit planning.
- Gathering of data.
- Importance of evidence.
- Rationalizing findings.
- Documenting and reporting findings.
- Audit follow up.
- Human relations issues surrounding the audit process.

It is not a bad idea to train a very large group to be internal quality auditors. Back-up people are then available if some individuals drop out, change assignments, or find they do not care for the activity.

FUNCTIONAL TRAINING

This second session of "department" training, carried out by departmental supervisors, should occur during the period surrounding D-Day and must be done no later than 30 days before the registration preassessment. Each session will run perhaps 90 to 120 minutes. The topics of this training are:

- Revisiting of the key facts taught in the orientation training (above).
- Review of the procedures, work instructions, and other documents that are the employees' guidelines.

Employees will find problems, flaws, and inconsistencies in the procedures and work instructions. These discoveries are to be expected and are highly desirable! Be sure to have the employees initiate Corrective Action Requests so that the problems they find can be addressed and fixed.

Finish the session by giving the employees the same quiz that was given at the end of the orientation training. This time, however, make it a "closed book" quiz. Go over their answers with them afterward.

79. Internally, who should be involved in the implementation process from the beginning?

CAPSULE ANSWER

Implementation begins with the senior manager and fans out to involve the operational and line managers as time goes on.

Implementing QS-9000 effectively is a top-down process. Commitment and involvement start at the top, and then spread out gradually over the weeks and months.

The first person involved must be the *senior manager* of the organization—the CEO, general manager, or equivalent officer. He or she will not actively run the implementation process in most companies, even very small ones. (However, this would not be a bad idea.) But the senior manager gives the green light, rallies the troops, provides resources, solves disputes, and maintains oversight to make sure the process continues to move along (Question 71).

Next involved is a group called the *Quality Steering Team*. Members include the senior manager plus the managers of each functional element of the company, as appropriate:

- Administration.
- Data processing.
- Design.
- Finance/Accounting.
- Manufacturing.
- Research and development.
- Sales and marketing.
- Service.
- Shipping and warehouse.
- Training/human resources.

This group is actively responsible for carrying out the approved implementation plan. It is trained in QS-9000 principles almost at the beginning of the project (Question 78). Its activities should be documented in a standard operating procedure, and records should be kept of its actions. It meets regularly (no less often than once per month, between launch and registration). Generally, its responsibilities include:

- Monitoring progress on the implementation plan, and making adjustments as needed.

- Reviewing and approving policy documents and standard operating procedures.
- Approving proposed solutions to gaps in compliance (especially where these cross over functional lines).
- Resolving operational disputes among departments and functions.
- Reviewing and approving corrective and preventive actions.
- Reviewing registrar proposals, and selecting the registrar.

The top functional manager in each area need not attend every meeting. Much of this work can be (and usually is) delegated. But the top functional managers should be actively involved in the process, to keep their subordinates from getting the wrong message.

The Quality Steering Team often stays intact as a permanent body after the implementation is complete. At the very least, its members usually surprise the group that meets at defined intervals for the management reviews (Question 31).

The "first among equals" in the Quality Steering Team is the *management representative* (MR). This is the person responsible to the senior manager for effective implementation and maintenance of the quality system. The duties of the MR are described in Question 75.

As the implementation proceeds, the next to be brought on board are the line managers, forepersons, supervisors, and leaders. (In very small companies, these may be the "senior managers" who sit on the Quality Steering Team.) Usually, their involvement begins with overview training and then intensifies as they start to review the quality system documents during the documentation phase.

Historically, this group is the most difficult to motivate. Of everyone involved, they are the most focused on short-term, deadline-related, production-oriented issues. They have little time for "theory" (to use a polite term), and often resent senior management's "inflicting" QS-9000 on them.

The question is often raised: "What's the easiest way to get middle management on board?" There is no "easiest" way. There is, in fact, no easy way at all. They are not going to identify anything as a direct benefit for a long time. To them, the endeavor will seem to bring only a lot of wasted time and useless paperwork. The best tactic here seems to be an enlightened mix of carrot and stick. The carrot is senior management's professed confidence that the QS-9000 process will help improve company performance, thereby benefiting all. The stick is that everyone is in deep trouble if implementation does not get done!

80. Should a consultant be hired to help implement a QS-9000 system?

CAPSULE ANSWER

It is not necessary—or necessarily desirable—to hire a consultant to help implement QS-9000. But if one is hired, be sure to choose a good one.

Not necessarily. In the long run, the most effective way to implement is without outside help. The process of trial and error and more trial and error results in a system that is truly unique to the business and fully owned by all levels and functions.

There is just one catch: The trial-and-error method takes a long time. A company pushed up against a customer's deadline may not have time to waste. Trial-and-error costs can be high. Cost-conscious companies will want to take the shortest possible route.

Some companies do choose to hire consultants to help them implement their QS-9000 systems. There is only one valid reason to hire someone: to shorten the time it takes to get the job done. Beyond that, any promises made by any consultant are a lot of hot air.

A *good* consultant can help get the implementation done faster, by using experience, creativity, and hard work to plan a system that fits the company and its needs while avoiding excess effort and mistakes. Although there is no evidence to support this, experience suggests that a reasonably priced consultant who knows his or her duties and works hard and conscientiously, will save a company money, compared with the trial-and-error approach. The consultant should most certainly reduce the level of misery and frustration.

What criteria identify a good consultant?

1. QS-9000 and ISO 9000 experience. There is no substitute for this background. A good consultant is one who has personally supervised an ISO 9000 and QS-9000 implementation not just once, but many times, from start to finish. This is no job for academics, self-promoters, or miracle workers. Look for someone with real-time scars.

2. Eclectic experience. A good consultant can be highly effective even if he or she does not have personal experience in the automotive industry. Look for a veteran in a wide range of industry sectors and company sizes. However, all other things being equal, if the choice is between two consultants who meet the first criterion (above), where one has personal experience in the automotive industry and the other one does not, choose the former.

3. Eclectic work experience. Some consultants have spent their whole careers as quality managers, or engineers, or in some other fairly narrow discipline. This is not a bad thing, in and of itself. But give a slight edge to a consultant who has worked in several areas of business.

4. Hands-on work ethic. Consultants can be divided into two general categories: those who talk and those who do. Hire a doer. A good consultant is not content to sit in a conference room all day, dispensing pearls of wisdom to the top brass. A good consultant deals not only with senior management, but also with the troops. He or she gets to know all the players by being alongside them on the production floor, training, writing or editing documents, cheerleading, cajoling, solving problems, moving the process along, constantly striving not only to implement the system but also to improve it.

5. Strong people skills. A good consultant is comfortable and effective dealing with people at all levels and functions of the company. It would be pleasant if he or she were universally liked, but this is not a requirement. The consultant must command respect and attention, even if (as is sometimes the case) the company elects not to heed his or her advice in certain instances.

6. Practical and down-to-earth approach. Every consultant has a general (and, preferably, effective) approach to any situation. A good consultant is practical, experienced, and flexible enough to adapt his or her approach to fit each client's situation and meet each client's needs precisely. A good consultant never forgets whose system it really is. Quick quiz: What is the sure sign of a consultant who has not a clue as to what he or she is talking about? A blind and arrogant demand that the client change to fit *his or her* system. Avoid this "my way or the highway" route.

7. Good personal fit. Companies are as different, and as individualistic, as people. By all means, pick a consultant who makes people feel personally comfortable. Work with the person will go on for a pretty intense and lengthy stretch of time. It will not always be pleasant, and it does not have to be. It only has to be effective.

81. Which elements of the system should be implemented first?

QS-9000 consists of:

CAPSULE ANSWER

The implementation process will address many issues and cover much ground. But several elements of the system should be addressed first.

■ Twenty ISO 9000-based requirements.
■ Three sector-specific requirements.
■ An array of "customer-specific requirements."

A company must comply with at least 18 of the first group, all three of the second group, and as many of the customer-based requirements as validly apply.

If the company is like most others, its process already meets perhaps 70 to 80 percent of the requirements. (This is especially true of audited suppliers to one of the Big 3.) The challenge, in implementation, is not to create an entire array of whole new systems where none previously existed. In large part, the challenge is:

■ To document and systematize the existing quality practices.
■ To implement those practices so that they are being carried out consistently.

Even so, the process typically lacks compliance to several of the requirements. There are, inevitably, many smaller gaps in compliance elsewhere in the system. Plus, a whole lot of top-down document review must be done, and new documents must be created. How does a company put together a plan for this? What is the order of priority?

There are several systems that should be created, documented, and implemented first. The reasons for their priority are:

1. They lay the groundwork for the others.
2. They tend to be among the most difficult and time-consuming to develop, debug, and implement.

QUALITY POLICY STATEMENT (REQUIREMENT 4.1—MANAGEMENT RESPONSIBILITY)

This statement has to be done first, and only senior management can do it. It must first be created and approved, and then communicated to the rest of the organization so that it is "understood, implemented, and maintained" (Question 81). The only way to get this done is to instill it into employees, over time, via repetition (Question 78). For that reason, it is a key part of both awareness training and orientation training.

But the main reason for its priority is that the Quality Policy Statement is meant to be the guiding light of the whole effort. Every facet of the quality system is aimed at fulfilling the Quality Policy Statement. Without it, the quality system is a collection of rules with no common purpose aside from "Management says to . . . ." The Quality Policy Statement must be meaningful and pertinent.

DOCUMENT AND DATA CONTROL (ELEMENT 4.5)

The standard operating procedure (SOP) written for compliance to this requirement tells how quality system documents are:

- Created.
- Reviewed.
- Approved.
- Distributed.
- Updated.
- Withdrawn when obsolete.

A company cannot issue any quality system document (not even the *Quality Manual*) until it has defined, in this procedure, how the document gains life, meaning, and clout.

The document and data control SOP belongs at the top of the document list. It must meet all the requirements of Element 4.5. (Don't forget to address the requirements pertaining to control of customer specifications as well as national and international standards.)

CORRECTIVE AND PREVENTIVE ACTION (ELEMENT 4.14)

This element of QS-9000 is the "problem-solving" element. It also helps drive the "continuous improvement" process, which is defined elsewhere (Question 27). It is *very* important that the corrective action process be defined and documented almost the instant the implementation process begins—for two reasons:

1. The process of reviewing and documenting existing quality practices will inevitably turn up many problems, gaps, and inconsistencies. These need to be addressed in a formal, disciplined manner, to ensure that responsibility is defined and progress is tracked.
2. For companies that have not had cross-functional problem-solving/corrective action systems, this element is one of the most difficult to implement. It takes time to educate employees as to how it works and what their responsibilities are. In some companies, it takes time for employees to understand and accept that this is a positive improvement process and not a negative "Let's write somebody up" process!

At the outset, define how the corrective and preventive action system is going to work. Document it in an SOP. Issue this (along with associated forms) as required in the document and data control SOP.

And then start promoting it like crazy. Encourage employees to document quality-related problems, of whatever nature, with Corrective Action Requests. This will help jump-start awareness. When employees see that the system really works, it will be evidence to them that QS-9000 is more than just a "paperwork exercise."

CONTROL OF QUALITY RECORDS (ELEMENT 4.16)

Of all the elements of QS-9000, at least fifteen specifically require that records be kept. Many of these are being kept already in one form or another. As SOPs are written, a company must identify the relevant records and define responsibilities, locations, and retention intervals. The records should be reasonably accessible and kept in an organized, protected manner.

Write a control of quality records SOP right away. Issue it in accordance with the document and data control SOP. Make sure everyone involved in the document-creation process is aware of the procedure, so that the required quality records are defined and handled in an approved manner.

82. Instead of creating documents from scratch, can existing documents be adapted?

Certainly. This is, in fact, a common approach. Possibilities include:

> **CAPSULE ANSWER**
>
> Adapting existing documents can be a good "shortcut" to doing the document writing job. But there are pitfalls.

- Where a similar organization can make its *Quality Manual* available, use it as a starting point or model.
- Documents already circulating within an organization may be adaptable as procedures, work instructions, and forms.

Adapting existing documents can save time and effort. But this approach is not a quick fix, and the following dangers are involved:

- It is unwise to simply lift existing procedures and work instructions and throw them into a QS-9000 system without thorough review. Procedures and work instructions may be in use already, but that does not mean they are right. It also does not mean that people are actually following them.
- It is risky to take someone else's procedures (or *Quality Manual*, for that matter), put a borrower's name on them, and put them into practice without thorough review. The company from which the documents have been acquired may seem "exactly like ours," but, rest assured, it is not. No two organizations are exactly the same.
- When companies adapt existing documents, they sometimes gloss over the text without fully understanding what it says. This can cause them to say things they do not intend. This often happens with the *Quality Manual*. There is a tendency to treat it as "boilerplate" and simply knock it out without understanding what the statements really mean.
- Organizations that adapt their own (or someone else's) document systems also run the risk of having systems that are too large, too complicated,

and too unwieldy. The mere fact that a document exists today does not mean it is worth having and using tomorrow.

To adapt existing documents for a QS-9000 system, do the following:

- *Preliminary:* Before any writing or adapting is done, define the format and structure for each type of quality system document. (This is a one-time exercise.) Each section of the *Quality Manual* should be structured the same way, with common headings and topical divisions. Similarly, define a standard format and structure for procedures and work instructions to follow.
- *Step 1: Decide whether the document really has to exist.* A document may exist today, but that does not mean it must be perpetuated. *Take a zero-based approach* to existing documents. Before starting the real work, subject each document to the Smell Test (Appendix D).
- *Step 2: Review related documents.* The Standard, the *Quality Manual*, and other related documents should be evaluated for their relevance.
- *Step 3: Identify the champion.* Each document (or, in the case of the *Quality Manual*, each section) should be assigned to a person who has sufficient knowledge (preferably, expertise) on the subject of the document. Being the champion does not mean that he or she does all the work. But the champion is the person who coordinates the efforts of the "doers" in reviewing and adapting a document.
- *Step 4: Identify team members.* These are the "doers"—the people who are actually doing the work that is being documented and who know best how the work is done. They also will be obliged to work in a way that is consistent with the finished document. They will be audited as to how well they are complying with the document, so they must buy into it. The best way to get their buy-in is to involve them, from the start, in the creation of the document.
- *Step 5: Edit the document.* With information obtained from the "doers," adapt the document so that it fits the company. If some company practices have to change as a result of this process, document them with a corrective Action Request (Question 66).
- *Step 6: Identify reference documents.* If specific reference documents or forms are needed to carry out the activity, or if other quality system documents are relevant, list them.
- *Step 7: Identify quality records.* If the activity generates a quality record, as defined or required by the Standard, list here the record's name, location, and owner.

- *Step 8: Get feedback from the team.* Give the draft to the team identified above. Have them read it, mark it up, and give it back. (It's very wise to give them a firm deadline for response!) Encourage them to be wholly honest. Is the document accurate? Is it workable? Can it be simpler? Does it need changes? How can it be made better?
- *Step 9: Create a second draft.* Use the feedback obtained above.
- *Step 10: Have a "dry run" done.* Give the document to the "doers" (the people who are obligated to follow it) and ask them to review it again. Have them work with it for a few days, under actual operating conditions. This is a more reality-based review process than the simple "read and edit" chore carried out earlier (Step 6). Instruct them to write down any thoughts they have on the document, and to return it by a specified date.
- *Step 11: Create a publication draft.* With the responses obtained in Step 10, create a publication draft. This draft will then be issued in accordance with the document control procedure.

At this point, is the job done? No. In fact, no quality system document is ever really done, not even the *Quality Manual.* As living documents, they are subject to change, update, and improvement as time passes and conditions change.

83. Are there any special steps for writing documents from scratch?

> **CAPSULE ANSWER**
>
> Many of the steps for creating documents from scratch are the same as those for adapting existing documents.

At least with the *Quality Manual,* there is an existing document to use as a model. But, with other documents, this may not be the case. And, with standard operating procedures (SOPs), it almost surely will not be the case. SOPs tend to be much more operation-specific than quality manuals.

Someone who has no model to adapt will have to create a document from scratch. This is not entirely a bad thing. Some people find the "from-scratch" process easier than the adaptation route (Question 82).

This method is specifically for SOPs, but it works just as well for quality manuals and work instructions. Follow this sequence:

- *First, do Steps 1 through 4 in Question 82.* They are:
  - Step 1: Decide whether the document really has to exist.
  - Step 2: Review related documents.
  - Step 3: Identify the champion.
  - Step 4: Identify team members.

In place of the previous Step 5, substitute:

- *Step 5A: Identify the process or processes.* The typical SOP may map out just one process, and a work instruction is almost always just one process. But some SOPs may cover more than one process. The task here is to identify, by name, the process(es) that the document will describe.
- *Step 5B: Define input and output.* For each process identified in Step 5A, define the "input" (or starting point) of the activity and the "output" (or result, or ending point) of the process.
- *Step 5C: Define the action steps.* Between input and output, action steps must take place in order to convert the input into the output. List those action steps in sequence. For each action step, document, as required:
  - Owner (name of function that does the work).
  - Whether a decision is required and, if so, what the criteria are.
  - Equipment and materials needed (most especially for work instructions).
  - Special skills needed, if any.
  - Whether inspection or some other verification is needed. If so, specify the workmanship standards or other acceptance criteria.
  - Whether the step is failure-prone or especially critical to meeting customers' quality requirements. If this is the case, more detail and precision may be needed to ensure strict consistency.
- *Step 5D: Create the initial draft.* Do not belabor this. Bang it out. The job here is not to "get it perfect" (for one thing, there is no way to do that); it is to get it done and into the hands of the "doers" for review and feedback.
- *Complete Steps 6 through 10 from Question 82:*
  - Step 6: Identify reference documents.
  - Step 7: Identify quality records.
  - Step 8: Get feedback from the team.
  - Step 9: Create a second draft.
  - Step 10: Have a "dry run" done.
  - Step 11: Create a publication draft.

Remember: A document job is never really done. A living document remains current to prevailing practice.

Besides—and this is the really depressing news—even though the people who really do the work have helped, serious "bugs" still remain in the documents. People who review the *Quality Manual*, SOPs, and so on, at this stage do not focus on them terribly well. They usually do not get around to this until the internal quality auditing process starts.

But at least something has emerged that is reasonably accurate and workable to start with. From this point on, keeping the documents current is up to the people who are obliged to follow them.

84. Does it make sense to adopt "canned" QS-9000 documents on a disk?

They cost money. Some of them cost *real* money. They can avoid some of the scut work of keyboarding, reduce the need for some basic decision making, and indicate a place to start. They can also help automate the document control process. But invest in them with care.

> ### CAPSULE ANSWER
>
> Canned QS-9000 "document systems" can cut out some of the work in creating and controlling the documents. But, as with everything, there are a few catches.

Here's why.

Many companies—some large, most very small—are marketing QS-9000 (and ISO 9000) "documents on disk." Some of these products are just text files with templates of quality manuals and procedures. Others include full-blown document control systems that are compatible with various types of PC networking programs. The idea is to offer a "head start" with the documentation phase of the QS-9000 program, and, at heart, the idea is good.

But such programs are often not sold on that limited premise. Their proponents tend to overpromise. Some go so far as to imply that the purchaser can simply plug the program in, crank out the documents, and *violà!* A very direct shortcut to registration opens up—a shortcut that more than justifies the price.

Don't count on that happening.

There is nothing wrong with automating the process of creating and controlling QS-9000 documents. The more that can be done without overburdening the implementation program, the better. One large multinational put its quality manual on line, on both its mainframe and its computer-aided design (CAD) system, on a read-only basis. That saved a lot of paper circulation, but it did not eliminate it entirely. And it did not help much with the subordinate documents (standard operating procedures and work instructions).

And, as explained in Question 82, there is nothing wrong with adapting QS-9000 documents from existing documents. These can be proprietary documents or documents provided by others. *But such documents are only a place to start.* In this business, there is no such thing as "one size fits all," or "plug and play," no matter what the salesperson says. The company still has to do the work:

■ Read and understand exactly what the document says.
■ Identify parts that may not be applicable.
■ Identify statements that do not fit the way the company works.
■ Amend the documents so that they:
  – Fit the way the company operates.
  – Use the company's own special language.
■ Review the result for compliance to the QS-9000 Standard.

What about the document control issue? Do these programs really help cut out some of the paper chase? Yes, *if* the company already has the PC (or mainframe) infrastructure to support it. Check out the programs' "system requirements" carefully. Are the appropriate PC networks and supporting software on board and well implemented already? Is the equipment strategically placed so that everyone who needs access to the QS-9000 documents already has access to a PC—and has received the training necessary to use it?

If not, *two* major implementations may have to be done at one time: the QS-9000 program, and the PC networking program. One at a time is probably enough trauma for most companies!

In other words, there is no such thing as a free lunch. Effective implementation is hard work and it must be done. Use whatever tools will help, but do so with open eyes.

85. What are the rules for creating effective quality system documentation?

CAPSULE ANSWER

By following the Ten Commandments, a company will have a document system that is lean, easy to understand, and useful to the employees who are required to comply with it.

The Standard does not prescribe any such rules. They are left strictly up to individual companies to devise. Long experience with implementing QS-9000 (and ISO 9000) systems has, however, taught that good, effective, easy-to-use document systems have certain common characteristics.

After all, these are not documents to be stuck in a notebook and thrust on some high shelf somewhere; nor are they reserved for the sole use of managers and auditors. They contain no state secrets. They describe and specify how the quality system works. They tell people what they must do to meet the quality obligations placed on them by the company.

They should therefore be clear, specific (to an extent), and easy to read and to use. What's the appropriate reading level? Sixth to eighth grade, at the absolute highest. What about people who don't read at that level? Create even simpler documents for them. What about non-English-speaking employees? Get them translations. Everyone must be included.

Here, then, are the strictly unofficial but virtually immutable Ten Commandments.

1. *Make each document prove that it must exist.* Often, existing documents will be adapted for use in the system. But the mere fact that a document exists today does not mean that it must exist and be included in the QS-9000 system. No document has an intrinsic right to exist. Perhaps a week, a month, or a decade ago someone thought the document was a good idea. It may not be a good idea now. Make a judgment on every single document. *Take a zero-based approach* to existing documents. Before starting the real work, subject each document to the Smell Test in Appendix D.

2. *Make each document consistent and compatible with other documents in the system.* This is a must. Look at the document hierarchy in Question 34.

- The *Quality Manual* must be consistent with the QS-9000 Standard (and must address all the "shalls" in the Standard).
- The standard operating procedures must be consistent and compatible with the *Quality Manual*.
- The work instructions must be consistent and compatible with the standard operating procedures (SOPs).

This means that the people doing the writing must have access to all the documents in the system. They must make it a practice to review related documents to ensure consistency and compatibility among them.

3. *Where possible, structure documents in the sequence of the activities being documented.* This is just common sense. Admittedly, it does not really apply to the *Quality Manual.* This policy-driven document (Question 35) responds to the requirements of the QS-9000 standard, so it really has no "sequence."

But SOPs and work instructions are another matter. They tend to have a sequence, because they describe how processes are carried out. It only makes sense to structure them in sequence, but, surprisingly, this does not often happen. Documents written out of sequence cause unending problems for users (and auditors).

4. *Where document text varies with actual practice, initiate a Corrective Action Request so that the practice is reviewed and changed.* As SOPs and work instructions are written, decisions to change the actual practices are often made, usually for two reasons: first, the existing practice does not comply with the Standard, and has to be changed; and second, management decides to change the practice to improve quality, efficiency, consistency, and so on.

Great. But just because a different practice is written down does not mean that people will respond in the way intended. To make sure that the practices themselves change, write up a Corrective Action Request citing the old practice and recommending the new practice, and assign it to the responsible manager in accordance with the system outlined in the SOP for corrective and preventive action. In this way, the change doesn't fall through a crack.

5. *Make documents operationally tolerant.* Too often, managers view the document-writing process as a chance to "tighten things up." They want to get super-specific and prescriptive about every single step. This is a mistake. Such tactics alienate employees and result in documents that are lengthy, bloated, difficult to read, and (ultimately) ignored.

When writing SOPs and work instructions, keep in mind the qualifications of the functions that are carrying out the activities. (These qualifications are written down. Review them during the writing [Questions 82 and 83].) Then write the documents so that they are only specific enough to ensure that the quality requirements of the activity are met. Rely on the experience, training, and good judgment of the employees to take care of the rest.

6. *Use functional titles, not personal names.* Remember that these documents must "keep up with the times." They must be accurate—always.

People come and go. Employees transfer, change jobs, get promoted. When personal names are used in the documents, they have to be revised, updated, and reissued each time an individual changes jobs. Unnecessary paperwork results.

Using functional titles is rather impersonal and non-people-oriented, which is unfortunate, but it saves a lot of work and headaches.

7. *Use present tense only.* These documents describe how activities are carried out *now.* They do not describe how activities will be carried out in the future. Some document writers adopt the prescriptive future tense legalese found in statutes or laws. Write what is happening now. Instead of:

The Sales Manager will sign each contract after review.

write:

The Sales Manager signs each contract after review.

Words and expressions like "will," "shall," or "is to be" are banned. Eliminating them has the additional merit of making sentences shorter.

8. *Make documents pithy by using short words, sentences, and paragraphs.* Short units are probably the single most important attribute of effective quality system documents.

- Strive always to find the clearest, most direct, and shortest way of saying things.
- Use everyday words; SOPs and work instructions are not opportunities to show off an elegant vocabulary.
- Keep sentences short. An average of nine words or less is highly desirable. Ban the use of the semi-colon (;). It has no place in these documents.
- Keep paragraphs skinny. Big blocks of type are hard to read. Use bullets and outlining styles.
- Documents tend to find their own length. A conscious effort to write short may still produce documents that motor on for a few pages. Try to break up long documents. An SOP that is longer than three or four pages is likely to be an SOP that is not read.

9. *Explain acronyms when first used.* Every company has its own jargon. In many firms, the jargon consists of acronyms. Some firms are more acronym-happy than others: entire conversations can be conducted in acronyms ("I did ACDUTRA at CNAVRES"). Have pity on the person who has to figure out what the clumps of capitals mean when sorting through the quality manual, SOPs, and work instructions.

"But doesn't everyone *know* what these things mean?" No, everyone does not. New employees don't, and they will be reading these documents. Neither do people from other areas of the company. Finally, there are the external (registration) auditors to think about. The better they understand the company's system, the smoother the audit will go. Count on it.

In each document, explain each acronym at the first usage. Afterward, use the acronym only. Readers will thank you.

10. *When in doubt, leave it out.* One of the main tasks will be to keep the entire system as lean as possible. The only way to do this is to be more vigilant about leaving things out than about putting things in.

86. When documentation is complete, what happens next?

> **CAPSULE ANSWER**
>
> "Finishing" the documents is certainly a milestone, but the documents are never really finished. This is when the real work begins.

When the *Quality Manual*, SOPs, and work instructions have been written, some people think they are done with the documentation.

They are not.

In principle, the QS-9000 documentation job is never finished. This is a living system. It must fit what is done and the way it is done. It must, therefore be flexible, adaptable to change. Processes undergo constant change as a result of:

- Customer requirements.
- Corrective actions.
- Employee suggestions.
- Continuous improvement activities.
- Strategic changes in the company.

As those changes occur, the documents must change also. (This is what makes document control so challenging. It is also what makes the document control element [4.5] a leading source of noncompliances during internal and external audits.)

No matter how thorough a company has been in developing the documents, they are probably only 70 percent "right." But why, after going through all the steps in Questions 82 and 83, and involving all the doers in the process of making these documents—why in the world would they not be right?

Simple. Many of the doers have not focused on the details in the documents. They skim them and scan them, nod, toss them back, and go on to the next thing. They do not yet fully appreciate that they will be *audited against the documents*. When they really focus on what the documents say, the changes come thick and fast.

That usually happens when the internal audit process starts (Question 88). By the time of the registration audit, the vast majority of the quality system documents will be in their second, third, or fourth revisions. In the meantime, there are several things a company should do to drive the implementation process:

1. Carry out orientation training. As described in Question 78, it is essential that line managers, supervisors, and leaders sit down with their people and review the relevant (and *only* the relevant) SOPs and work instructions with them, get their feedback, and encourage them to suggest changes.

2. Talk up the system. Make employees understand what is happening. Try these lines:
   - "This is not just a fad or a phase."
   - "We work to this system now."
   - "We will be audited to make sure we do."
   - "You have a role, you can influence it, but this is not just for now, it is for good."

3. Stress that the QS-9000 system is not meant to stifle initiative or innovation. The company follows the system but is also required to improve it. Management needs to communicate two messages here, consistently:
   - "What does the procedure say? Are we following it?"
   - "Does the procedure reflect the best practice? How can we improve the system?"

4. Make the documents a focus of problem-solving activities. When a nonconformity occurs—a customer complaint, a product defect, whatever—and the chase is on to find the root cause, find out first whether the documented procedure was followed:
   - If the procedure was followed and the nonconformity occurred, how should the procedure be changed?
   - If the procedure was not followed, would additional training solve the problem?
   - If there was no procedure covering the activity, should one be implemented? (Be careful here. Addressing problems simply by writing documents is what leads to the dreaded "document bloat.")

In other words, *work the system*. Every day, all the time. And see to it that everyone else does also.

87. What's the best way to set up an internal audit program?

> **CAPSULE ANSWER**
>
> An audit team must be multilevel, cross-functional, proactive, and effectively trained.

If there is a single secret to successful implementation, it is a good strong internal audit program. Timing is important, too. The audit team should be trained and ready to go at just about the exact time when the first SOPs are written, approved, and issued.

To make sure that the audit program contributes as it should to the implementation process, care must be taken in the selection, training, and initiation of internal quality auditors.

SELECTING THE AUDIT TEAM

The Standard says that internal auditors must be "independent" of the management of the area they are auditing. This is restrictive, but not overly so. There are many ways to structure an audit team without violating this rule.

Some companies have appointed people to be "dedicated," full-time internal quality auditors. But this smacks of the confrontational "Inspector Gadget" approach that is, for the most part, in the past, and should stay there. Other companies have gone so far as to hire outsiders to do their internal audits for them. Despite the pretty glaring contradiction—there is more than a slight technical difference between "internal" and "external"—some registrars have actually approved this approach.

But the internal process is key to successful implementation. If properly selected, the internal audit team becomes a band of "ambassadors" of the quality system to the rest of the organization. They carry the torch, spread the word, become authorities on the subject. For those reasons, the internal audit team has the following ideal profile:

■ *Multilevel.* Some companies more or less automatically assume that internal quality auditors have to be management or supervisory people. This is neither true nor desirable. To appoint only managers or supervisors to be internal quality auditors is to say to the workforce: "QS-9000 is a management system—us against you." This is exactly the *wrong* message. The system belongs to everyone.

- *Multifunction.* Some companies more or less automatically assume that internal quality auditors must be "office people"—sales, engineering, or human resource types. This is also not true and not desirable. Warehouse, production, and maintenance people can be and have been excellent internal quality auditors. At least one representative from each function of the company should be appointed to the internal audit team.

- *Proactive.* Every manager worth his or her salt knows the difference between an employee who is there for the check, and an employee who is interested in the job, the company, and the future. Make sure the internal auditor candidates fit the latter profile.

Internal auditing is not rocket science. No high amount of education is required. The auditors need to be able to read and write, ask reasonably intelligent questions, follow through, and manage themselves.

During the selection process, put out the word that anyone who would like to volunteer for internal quality audit duty is welcome. This will help in the search for people who are otherwise qualified and interested but have perhaps been overlooked.

A common question is: How many people should be on an internal audit team? According to a rule of thumb that has worked in many places, about 10 percent of the total head count should be trained as internal quality auditors. This ratio tends to produce a team large enough to handle the audit load comfortably, while providing for extras in case people drop out.

TRAINING THE INTERNAL AUDIT TEAM

Like all other employees, internal quality auditors must have appropriate training. Many training programs are available. (The Automotive Industry Action Group (AIAG) has announced that it is offering a "sanctioned" QS-9000 internal auditor training course. Contact AIAG for more information.)

The typical program runs two or three days and is based on the ISO 10011 "Guidelines for Auditing Quality Systems." The best courses also spend time on the human relations aspect of internal quality auditing. They stress, again and again, that the audit program must never become confrontational. Audits must be thorough and persistent, but also fair, objective, and impartial. And audit findings—both positive and negative—must be supported by objective evidence.

A typical outline for a two-day internal audit course looks like this:

QS-9000 Requirements

Internal Quality Auditing
    Importance of internal quality auditing
    Purpose of internal quality auditing
    Principles of effective internal quality auditing

The Nine Phases of an Internal Quality Audit
    Phase 1: The Audit Schedule
    Phase 2: Auditor Assignment
    Phase 3: Checklist Preparation Evidence
    Guidelines for Preparing Checklists
    Phase 4: Opening Meeting
    Phase 5: Gather Information and Evidence
        The Effective Auditor
        Auditee Tactics
    Phase 6: Handling Suspected Noncompliances
    Phase 7: Rationalize Findings
        Types of Findings
    Phase 8: Document Findings
    Phase 9: Closing Meeting

Follow-up

INITIATING INTERNAL QUALITY AUDITORS

The final phase of training should be positioned as a "practice audit." Group the auditors in pairs and assign each pair a procedure to audit. Do this the day after training ends. Have them go through all the steps of the audit. Then convene the entire team to meet and review the results.

Once the entire team has completed a practice audit, they are ready to audit "for real."

88. How can the effectiveness of an internal quality audit program be maximized?

CAPSULE ANSWER

Effective internal auditing is a matter of prudent scheduling, nonconfrontational tone, thorough documentation of findings, and appropriate follow-up.

Internal quality auditing is one of the three "reinforcement mechanisms" of QS-9000 (Question 13)—the functions of the system that keep it active, contributing, and improving. The other two are the management reviews required by the Standard (Question 31) and the surveillance audits carried out twice a year by a registration body.

Why are the internal quality audits very important?

- They are management's window into the quality system. They give management an objective, evidence-supported view of the status and effectiveness of the quality system.
- They educate employees (auditors as well as auditees) about the quality system and the functioning of the company.
- Most important, at least in the beginning, they drive the implementation process. Most employees do not just drop the way they are doing things and hurl themselves into the arms of the QS-9000 system. Many are indifferent at best, and some are downright skeptical. Internal quality audits are evidence to them, over time, that the system is real, important, and here to stay.

When it is so important, how can the effectiveness of the internal audit system be ensured?

For starters, select the audit team with care. As discussed in more detail in Question 32, it is best to have an audit team that represents a good cross-section of the levels and functions of the company. Do not restrict the audit team to quality people or supervisors.

Here are some other ideas for maximizing the effectiveness of the internal audit program.

SCHEDULING

A company is required to audit all the processes and procedures implemented as part of the QS-9000 system. That is a fairly large "pie" to divide up. But if each individual audit drags on for a long time, it cuts into the lives of both the auditees and the auditors. On average, each audit should take no more than eight hours, from preparation and data gathering through reporting, when done by a two-person audit team.

What is the best way to set up the audit schedule? Arrange it by:

■ Clause of the Standard (4.1, 4.2, etc.).
■ Area (Department) of the company.
■ Procedure.

When the audit is done by clause, each audit will cover at least one procedure (some, probably, more than one). Each audit will also cover, potentially, more than one area of the company. That can be a lot of ground to cover.

When the audit is done by area or department, more than one procedure will most likely be audited at a time. Take a look at the Procedure/Department Chart in Appendix E. For some departments, such as production, several large procedures may have to be covered. Once again, the result can be very lengthy audits that take up lots of both the auditor's and the auditee's time.

Probably the most practical approach is to arrange the audit schedule by procedure. This divides the system into fairly digestible chunks. Some procedures, as shown in the Procedure/Department Chart, cover more than one department or area of the company. Those audits may take longer, even on a sampling basis.

When an audit is done this way, be sure to include, in the scope of all audits, the following:

■ Quality policy awareness. This covers the requirement under Element 4.1 (Question 22).
■ Effective implementation of procedures. This covers the requirement under Element 4.2. Procedures may be written and in place, but if they are not being followed, then this is a noncompliance under 4.2.2b.
■ Work environment. Element 4.17 of the Standard requires that the "work environment" be evaluated as part of each audit. Issues such as housekeeping, cleanliness, temperature, and so on, are included.

The entire system must be audited at least once before registration. So, from D-Day on—the period referred to as "Crunch Time" in the Implementation Timeline (Appendix B)—intensive auditing will be ongoing. This is good for both auditors and auditees. For auditors, it is good training;

for auditees, it is good practice. More important, it teaches them about the system in a way that nothing else can.

Once a company is registered, it should, at a bare minimum, audit the entire system at least once per year. Twice tends to be the norm. Three times is not unheard of; actually, it is recommended for the first year or two, until the system is mature and reaches steady state.

"Problem areas" will need to be audited more often than others. The Standard requires that the audit schedule be driven in part by audit results, which only makes sense. Registration assessors check for this response.

EFFECTIVE AUDIT PRACTICES

- Publish the schedule well in advance. It is a good idea to schedule audits for "the week of" rather than for any particular day. Then let the audit team and the auditees establish the day and time that fit their schedules best. But take care to enforce the schedule. Do not let it slip. Once it gets behind, it is very hard to catch up. And letting it slip sets a bad precedent.
- While an audit team is new, let it audit the same area a couple of times. Then, start to rotate audit teams among different areas of the system. It is an error to let the same team audit the same area over a long period of time. Inevitably, objectivity starts to suffer.
- Middle management can sometimes be a problem. As the segment of the management team that is under the most pressure—or at least perceives itself that way—middle managers sometimes resent the time they must devote to being audited. They also resent the time it takes for their employees to do audits. Be alert to this reaction and deal with it directly, during the initial stages of the implementation.
- Auditing must never become an adversarial process. Some antagonism cannot be avoided. But when it occurs, it needs to be addressed and eliminated.
- In large part, it is up to the auditors, not the auditees, to set the tone. They must approach their work with an attitude of fair and objective persistence. Ban the following terms from auditors' lexicon:
  - Writing you up.
  - Deficiencies.
  - Violations.
- Be sure that auditors audit not only compliance with the system (the usual objective) but also effectiveness of the system. This requires some thought and insight, especially at the reporting stage.

■ Make sure the auditors support their findings with documented, objective evidence. This is as important for positive findings as for negative ones. Failing to gather objective evidence on a consistent basis can result in noncompliances from registration assessors. Even worse, it can, over time, erode the credibility of the findings and, therefore, of the program itself.

FOLLOW-THROUGH

■ Handle internal audit noncompliances as part of the corrective action system (documented in the procedure for corrective and preventive action).

■ From time to time, analyze audit results by department and by clause of the Standard. Report these analyses at management reviews. The reports give solid evidence of the effectiveness of the system.

■ When an area of the system (a department, or a quality system function) turns up chronic noncompliances:
  – Supplement corrective actions with follow-up audits, carried out by the same audit team.
  – Increase the frequency of audits, at least in the short term.
  – Highlight these issues at management review meetings.

GENERAL

The audit team is a valuable asset to the company. They are performing an essential service, and it is vital that they be recognized for it.

■ Hold regular meetings so that they can air their experiences (or grievances).

■ Consider treating them to lunch or dinner several times a year, as a way of thanking them.

89. What are some of the common perils and pitfalls to effective implementation?

Just as every company is different, so is every implementation. But there are some common "pitfalls" that a company should do its best to avoid:

> **CAPSULE ANSWER**
>
> In the end, successful implementation requires an approach that is persistent and proactive, rather than defensive and resentful.

- *Trying to implement "from the bottom up."* It is surprising how often this happens. Sometimes, top management is direct about it: "Leave us out until the very end." Even more insidious is the senior management group that appears to be involved but is not. The rest of the workforce reads this loud and clear, and it can add months to the implementation process. *Top management must lead and drive the process, consistently and persistently.*
- *Trying to do "just enough to get registered."* Some companies say, "Let's just do QS-9000 for Process A and B, or Plant D and E, because those are the only ones that ship to the Big 3." Such half-measures are, in the end, self-defeating. Going about things in this way eats up time and resources. Successful QS-9000 implementation spans the entire organization. It brings everyone in.
- *The Lone Ranger syndrome.* Project champions, such as management representatives, sometimes take too much of the work upon themselves. Successful MRs learn to delegate, by selling, cheerleading, wheedling, and cajoling if necessary. Getting implemented and getting registered require 150 percent effort by *all* concerned. Everyone will share in the benefits, so everyone should share in the pain.
- *Getting carried away with documentation.* QS-9000 is, in part, about documentation. But another key to success is to create only as much documentation as is necessary to meet the Standard—plus documentation felt to be meaningful and value-added. Projects have been torpedoed by paperwork bloat. Use the Smell Test in Appendix D. Be ruthless and vigilant.
- *Springing the system on the workforce all at once.* QS-9000 is not a state secret. Its success, in large part, depends on the acceptance and involvement of everyone in the company whose work affects quality. Plan for the implementation with ongoing awareness sessions (Question 78). In this way, as the project gains momentum, people find themselves working with it instead of around it.

■ *Failure by champions to become educated in the requirements.* No one is claiming that reading the QS-9000 requirements is fun. It is more like undergoing root canal work. But there is no substitute—not this book, nor any other book—for reading them and understanding them. The requirements cannot be learned from one read-through. They can't even be learned by studying them for hours on end. Learning the requirements calls for repeated exposure.

■ *Regarding the QS-9000 system as separate from the rest of the business.* It will seem separate at first. But part of effective implementation is making the system "part of how things work" rather than "this program we have to do for GM." The successful, contributing QS-9000 system— that is, a benefit, rather than an expense—is one that becomes transparent within the company.

■ *Considering the work "done."* A QS-9000 system is not a "thing." It is a constantly evolving process, an evolution, a way of doing business that changes and adapts as your business must change and adapt in order to survive.

PART SIX

Registration

90. What are the general requirements for registration to QS-9000?

Basically, any company may register to QS-9000. Some are required to. These are companies that supply, directly to Chrysler, Ford, or GM:

■ Production materials.
■ Production or service parts.
■ Certain services (heat treating, painting, plating).

CAPSULE ANSWER

To register to QS-9000, a company must meet all the requirements applicable to its location, pass an audit by an accredited registrar, and undergo semiannual surveillance assessments.

All other things being equal, a company should not register to QS-9000 unless it is required to by a customer (a Big 3 location, or a direct supplier to the Big 3). QS-9000 is aimed specifically at the types of companies and processes described above. Other types of companies (service providers, for example) can register, but they must meet all the requirements, which can be a torturous process. Rather than registering to QS-9000, such companies should register to ISO 9000.

To register to QS-9000, a company must:

1. Implement a system that meets all the requirements in *Quality System Requirements QS-9000*, Parts I and II. The only elements that may be deemed "nonapplicable" are:
    ■ Design Control (Element 4.4), for locations that are not design-responsible.

- ■ Servicing (Element 4.19), for locations not contractually required to provide servicing to customers.
2. Meet the "customer-specific" requirements in Part III, for the customers (Chrysler, Ford, GM) that are served.
3. Operate the system for three, and preferably six, months.
4. Complete at least one cycle of internal quality audits (Question 32).
5. Undergo an audit of its organization, based on *Quality System Assessment* (QSA), by a registrar accredited by an approved accreditation body (Question 16).
6. Resolve and close out all noncompliances raised by the audit.
7. Undergo surveillance assessments every six months, and close out noncompliances raised by those assessments as required by the registrar.

91. How should a company go about choosing a QS-9000 registrar?

> **CAPSULE ANSWER**
>
> Registration bodies are not all the same. Choose carefully.

There are at least 60 quality system registrars operating in the United States as of mid-1996. Others, knowing a growing market when they see one, are entering the business every day. How does a company choose?

Fortunately, the "first cut" has already been made. A company must select a registrar that is "accredited" to issue QS-9000 certificates by a Big 3-sanctioned accreditation body (Question 20). Accreditation means that the registrar:

- ■ Follows the international code of practice for quality system registrars (EN 45012).
- ■ Follows the QS-9000 "Code of Practice for Quality System Registrars."
- ■ Has been assessed and approved on these and other technical issues.

Those requirements bring the list down to about 24 accredited registrars with North American locations (see below and Appendix A). All of these registrars are "certified" as being competent, experienced, and ethical. A list is available from the Automotive Industry Action Group (AIAG). Another list appears in Appendix A.

To pick the right one from this bunch, should a company shop price and go for it? Not exactly.

Price is certainly a factor. But experience shows that several other factors are just as important, if not more so. When evaluating potential registrars, consider the following:

1. Compatibility. Registration bodies are companies. Every company has its own culture, language, demeanor, "feel." Some are relatively relaxed and easygoing. Others are stuffy, bureaucratic, and uncommunicative. Some bend over backward to hold their customers' hands. Others have a remarkable ability to make the registration process more difficult than it needs to be. Here are some tips:

   ■ When narrowing the selection, interview at least one person (preferably an actual auditor, not just a "business development" person) by phone or in person. Size up that person. Judge for yourself the "fit" of cultures of the two companies.

   ■ Word may be out to the effect that certain registrars are "easier" in audits than others. Do not let this gossip affect the selection process. The gossip is most likely not true. And even if it is, don't seek out an "easy" registrar. Look for one who will give a fair, objective, rigorous assessment, initially and ongoing. Otherwise, the fee is not buying full value.

   ■ By all means, pick a registrar whose culture is compatible. After all, the company will be "married" to the registrar for at least the first three years.

2. Name recognition. Some of the registrars on any list have "household" names, at least in the consumer sense. Underwriters Laboratories is probably the best example. Name recognition, in the general public sense, is not terribly critical. What's important is that the registrar's name must mean something to the company's customer(s)—not the Big 3 customers, who are bound to accept QS-9000 registration from any registrar accredited by a Big 3-recognized accreditation body, but the non-Big 3 customers. When selecting a registrar, look beyond today's needs.

   ■ If a company is doing business overseas, or plans to, its customers and/or prospects should be polled. Get an idea from them about the registration bodies they are familiar with. Lean toward a registrar that has name recognition in the areas and market sectors with which the company does business, or plans to do business. This may mean selecting a registrar that is accredited not just by the Registrar Accreditation Board (RAB), a U.S. accreditation body, but also by one of the overseas bodies.

3. Location. Increasingly, registrars' operations have North American bases. Some even have multiple branches around the country—a benefit to customers because of saved travel costs. Even so, location is still an issue. The farther the distance from a registrar (and, even more important, from an audit team), the higher the costs. (Don't be unduly swayed by registrars that "include" travel expenses in their quotations. The cost is still in there, under other headings.)

■ When evaluating registrars, find out where their assessors are based. Some may employ field people who work out of their homes, perhaps very close to where the assessment must be done. Others may bring people in from a great distance. Wherever they're from, the cost of getting them to the assessment and back is billable, and these costs are not necessarily foreseeable.

4. Cost. Some clients do not realize that there is no set fee or expense structure for quality system registrars. The process is totally market-driven. The costs, for the same facility, can swing up to 100 percent among five or six different registrars!

■ Obtain specific and detailed quotations from a number of registrar candidates. (Six is a good number.) Make sure each quotation includes at least the following:
  – Daily rate for assessment (registration and surveillance).
  – Estimated number of days for registration assessment.
  – Estimated number of days for surveillance assessment.
  – Expenses reimbursable by the company being assessed.
  – Location of assessor(s).
  – Cost of registration renewal (if any).
  – Cost of document review.
  – Cost of certificates (some registrars charge for extra ones).
  – Cost of additional accreditations (with RAB, the first one is free; others registrars may charge for all).
  – Administrative fees.
  – Application fees.
  – Other fees.

■ Line up and analyze the overall expenses, including everything, *over the first three years of the relationship* (the typical length of a "registration").

■ Pay special attention to the registrars' policies on travel expenses. Some actually charge for travel *time* as well as costs.

■ Ask whether the registrar carries out a complete systems reaudit to "renew" the certificate. Most do. Some do not. This is a major cost factor to consider.

- Reduce all the quotations to a single amount per registrar, and then compare them.

5. **Industry experience.** How much experience does the registrar have in the company's particular market sector? All registrars accredited by Big 3-recognized accreditation bodies are qualified to do QS-9000 audits. Accreditation includes some presumption of experience. Each audit team is required to have at least one member who has "relevant experience" in the automotive industry, but it's a good idea to select a registrar that has actually registered companies in the company's particular market niche. More experience and a better understanding of the process and the market tend to result in more effective audit results—and audits that are more efficient. Less company time will be spent "educating" the assessors.

A company that does business overseas, as well as in North America, may want to choose a registrar that is accredited not only by the U.S. group (RAB), but also by one of the international accreditation bodies (Question 20).

TECHNICAL REQUIREMENTS

1. A company must select a registrar that is accredited to issue QS-9000 certificates by an accreditation body recognized by the Big 3.

TECHNICAL GUIDELINES

As noted above, Appendix A is a list of accredited QS-9000 registration bodies.

92. What are the steps involved in obtaining QS-9000 registration?

CAPSULE ANSWER

Although practices may vary slightly among registrars, the registration process itself is well defined and takes place in a series of steps.

The registration process is fairly well defined in the *QS-9000 Requirements* and *Quality System Assessment* documents. Even so, there may be some variation from registrar to registrar. They are, after all, independent entities. They are free to set their own policies, as long as those policies, as interpreted by their accreditation body (Question 20), meet the relevant requirements.

The process described below is general. Variation in the process is common and depends in large part on:

- Registrar policies.
- Size and activities of a company.
- Scope of the company's process.
- Whether any of the company's sites are design-responsible (Question 45).

STEP 1: IMPLEMENT QS-9000

Assuming that a company will implement QS-9000 first, its system must be fully implemented and up and running for a minimum of three, and preferably six, months prior to the registration audit. This is how long it take to build up the amount of evidence, records, and so on, that assessors expect to see.

For details on implementation, see Question 74. In capsule form, implementation means:

- All required elements of QS-9000 are in operation.
- At least one complete cycle of internal quality audits (Question 32) has been completed.
- At least one management review (Question 31) has been completed.
- Corrective and preventive action activity (Question 66) is ongoing.

STEP 2: SELECT A REGISTRATION BODY (QUESTION 91)

Begin this process while implementation is going on. Have a registrar selected, on board, and scheduled no later than three months prior to the target registration date. (Registrars' schedules vary; some may need more lead time, depending on market conditions.) Allow time to learn and digest the particular registrar's policies and procedures. Here are some points to remember:

■ Most registrars offer helpful general training in QS-9000 and ISO 9000.

■ A registrar will answer questions about interpreting the Standard. This too can be helpful.

■ If a company has multiple sites, coupled perhaps with some design responsibilities here and there, the registrar can answer questions about how to structure the system.

■ Registrars are not allowed to "consult." They will not tell how to implement a system, how to fix what's wrong, or how to develop systems that will meet the requirements.

STEP 3: OBTAIN A PREASSESSMENT (READINESS REVIEW)

This is an optional step and it costs money, but it is highly recommended.

A preassessment is a sort of "practice audit." Its results do not count. Its scope usually does not include an entire quality system. The company and the registrar determine what will be audited and how long that audit will take.

What is the point? A preassessment gives a taste of what a QS-9000 audit is like. (Very important. Employees who have been through other supplier audits may think they know what's coming, but odds are they do not.) It gives some feedback on how the system is doing, so that it can be fine-tuned before the registration audit. Some advice is in order:

■ Unlike the registration audit, in a preassessment, the registrar can usually be directed to the areas that the company wants covered. *Always* have a preassessment of the parts of the system about which the company is *least* confident.

■ In addition to the above, *always* make sure the preassessment covers the following areas of the system:
  – Management review.
  – Corrective and preventive action.
  – Internal quality auditing.

STEP 4: UNDERGO A "DESKTOP STUDY" OF THE QUALITY SYSTEM

This study may be done prior to preassessment, depending on the registrar's policies. The specific documents can vary from registrar to registrar. Some only want the *Quality Manual* (Tier 1) (Question 35) in advance. Some want to see the standard operating procedures (Tier 2) (Question 36) also. And some will not want to see these documents "off site." Instead, they will review them as the first part of the on-site assessment (see below).

Virtually every registrar does a desktop study. It may consist of reviewing the *Quality Manual*, comparing it with the requirements of QS-9000, and assessing it to make sure that the *Quality Manual* complies with the requirements.

If a company has done a thorough and conscientious job of developing its quality system, the desktop study should pose no problems. There will always be a comment or two, a niggle here and there. Odds are the registrar will find at least one important issue to address. Address any issues by either changing the documents or explaining the company's position to the registrar.

STEP 5: UNDERGO AN ON-SITE AUDIT

This is where the rubber meets the road. White-knuckle time. The registrar's audit team shows up on the appointed day, checklists in hand. If they did the desktop study in advance, they have already determined that the quality system, as documented in the *Quality Manual*, conforms to the Standard. Now they want to see whether the company is actually doing what the manual and related documents say it is doing.

The audit team usually consists of two or more people, depending on the size of the facility and the scope of the process. The typical personnel are:

- Lead auditor, who handles relations between the audit team and the auditee personnel. This person plans the audit, supervises the auditors, and takes the lead on interpreting the Standard and the audit results.
- One or more additional auditors. These are qualified but usually somewhat less experienced people. The audit team must always include someone with "relevant" automotive industry experience. Most registrars try to include a person whose experience in the field is even more specific.

The auditors, following a predetermined plan and checklist, audit the system (Question 95). At the end, they give a verbal report of their findings. Any noncompliances that they have found are delivered in writing, with supporting evidence (Question 96).

STEP 6: CORRECT AND CLOSE OUT NONCOMPLIANCES

All noncompliances reported during a registration audit must be closed out before a registration certificate can be issued. This applies even to a report of only one trivial noncompliance (and most auditees rack up a few more than that). A company is not permitted to say or claim that it "passed" or "got registered" until all noncompliances are corrected, closed out, and confirmed by the registration body.

Most often, the registrar will agree on a period of time (usually 90 days) to correct any noncompliances. It will also indicate which ones require on-site verification. Other noncompliances may be closed out via the submission, review, and approval of revised documents or procedures.

STEP 7: RECEIVE THE REGISTRATION AND THE CERTIFICATE

Registration is conferred when the registrar issues a certificate of registration. This includes the "scope of registration," which describes exactly what processes and locations are covered by the registration.

When a company is registered, is it done? Is it time to relax? Can things go back to "the way they were before?" See Question 99.

93. What are the rules governing QS-9000 audits?

CAPSULE ANSWER

QS-9000 has specific requirements governing registration audits. Various guidance documents offer pointers for implementing effective audit programs of other kinds.

QS-9000 quality system audits—most especially, the registration audit—are required to answer three questions:

1. Are all processes defined and documented?
2. Are all processes deployed and implemented as documented?
3. Are all processes effective in yielding expected results?

The QS-9000 documents include some specific requirements for how registration audits must be carried out. The chief document for this is *Quality System Assessment* (QSA). The Standard also prescribes the number of audit days to be utilized, based on the number of employees in the facility (see below).

ISO also publishes a series of guidance documents covering quality system audits: ISO-10011: Guidance for Auditing Quality Systems.

These guidelines are not mandatory for firms seeking QS-9000 registration. But their guidance is well worth consulting while setting up the internal audit programs required by the Standard (QS-9000, Element 4.17).

TECHNICAL REQUIREMENTS

1. Registration audit teams must include at least one member who has "relevant" automotive industry experience.
2. The audit must encompass all QS-9000 requirements:
   - Part I (Elements 4.1 through 4.20)—except, when not applicable, Elements 4.4 (Design Control) and 4.19 (Servicing).
   - Part II (Sector-Specific Requirements).
   - Part III (Customer-Specific Requirements), as relevant.
3. Registration audit checklists must include all entries in the *Quality System Assessment* (QSA).
4. Audits must include all elements of the quality system implemented to meet the needs of automotive customers. Elements that go beyond the QS-9000 requirements are included.

5. The registration audit must evaluate all quality system elements for:
   - Effective implementation of requirements (Are the requirements being met?).
   - Effectiveness in practice (Are the practices effective in meeting the requirements?).
6. Registration audits must review:
   - Customer complaints and company responses.
   - Results and actions related to:
     - Internal quality audits.
     - Management reviews.
     - Progress toward continuous improvement targets.
7. The audit team must provide a full audit report. The report must include opportunities for improvement.
8. Required registration audit days (August 25, 1995, revision) are:

Number of Employees at Site	On-Site Audit Days Required	
	Registration Audit	Surveillance Audit
1–15	2	1
16–30	4	1
31–60	5	1.5
61–100	6	1.5
101–250	8	2
251–500	10	2.5
501–1,000	12	3
1,001–2,000	15	3.5
2,001–4,000	18	4.5
4,001–8,000	21	5.5

TECHNICAL GUIDELINES

ISO 10011-1

ISO 10011-1 (1990): *Auditing* is a document that defines the quality system audit process. Its guidelines can be applied to all three types of quality system audits:

1. Internal audits (Question 32), which are required for firms registered to QS-9000.
2. External audits, in which a customer audits a supplier (something a company may well do as part of its conformance to the Standard's "Purchasing" requirements (QS-9000, Element 4.6)).
3. Extrinsic audits, conducted by third parties to assess a firm's eligibility for QS-9000 registration.

> **APPLYING THE STANDARD**
>
> Quality system auditors do not content themselves with interviewing managers in conference rooms. They roam the working areas and insist on interviewing employees and reviewing working documents.

Part 1 of ISO 10011-1, *Auditing* gives detailed guidelines for conducting quality system audits. For example, it defines the various personnel:

- *Auditee*—the organization being audited.
- *Client*—the organization requesting the audit. In internal quality audits, the client may be company management or the management representative. In registration audits, the client is the registration body.

The document also describes the *audit plan* (the audit objectives, personnel, language, and logistics) and the *working documents* normally used in quality system audits:

- *Checklists*—for auditors' guidance.
- *Forms*—for recording audit observations (see below) and for documenting evidence of findings.

> **APPLYING THE STANDARD**
>
> Some claim that the real rules of quality system auditing are:
> 1. If it moves, train it.
> 2. If it doesn't move, calibrate it.
> 3. If it's not written down, it never happened.

The audit process detailed in ISO 10011-1 includes:

1. The *opening meeting,* in which the audit team and auditee personnel:
   - Become introduced.
   - Agree on the scope and objectives of the audit.
   - Discuss audit methods.
   - Agree on the audit logistics to be employed, including resources, facilities, and guides.
   - Confirm the time and place of the closing meeting.
2. The *examination,* during which auditors roam the auditee's area, guided by their checklists. They are escorted by a representative of the auditee, whose task is mainly to provide them with necessary access. During the examination, auditors *collect evidence* concerning conformity and nonconformity, through:
   - Observing operations.
   - Interviewing employees.
   - Examining working documents and records.
3. During the audit, auditors gather *observations* (documented with objective evidence) about the state of the quality system. These observations are not necessarily nonconformities. Decisions on nonconformities are not made until the audit team's meeting, held just before the closing meeting. Auditors do, however, point out nonconformities to their escorts upon noticing them, without making judgments. Before the closing meeting, the audit team meets to evaluate its findings. The lead auditor makes the final decisions about the *nonconformities* (Question 96) to be reported to the client.

> **APPLYING THE STANDARD**
>
> A locked cabinet or cupboard is an invitation to a quality system auditor's curiosity.

At the closing meeting, the audit team presents its findings. These include whatever nonconformities were found during the audit, backed up with supporting, objective evidence. Usually, the auditee management is asked to:

- Sign the nonconformity reports, acknowledging them.
- Provide a deadline, by which date corrective actions will have been implemented.

The audit team does *not* make recommendations about corrective actions. Its task is to point out and document nonconformities to the Standard. It is up to the auditee to devise appropriate remedies.

ISO 10011-3

ISO 10011-3 (1991): *Management of Audit Programs* provides guidance for the ongoing management of quality system audit programs. It discusses:

1. The suitability of audit team members.
2. Monitoring and maintenance of auditor performance.
3. Management of audit programs.

Like ISO 10011-1, ISO 10011-3 is well worth consulting while setting up internal quality audit programs.

94. What qualifications must ISO 9000 auditors possess?

> **CAPSULE ANSWER**
>
> Registration auditors must meet general qualification requirements set by accreditation bodies. They must also receive specific QS-9000 training.

The Standard mandates qualifications for both internal quality auditors and "extrinsic" (registration) auditors. According to the Standard, quality system audits must be "carried out by personnel independent of those having direct responsibility for the activity being audited."

People performing registration audits must also meet this requirement. QS-9000 is very specific in the area of "independence." No organization that has provided consulting to a client since August 1994 may act as that client's registrar, nor may that company supply auditors.

In addition, registration auditors must meet the qualifications of the appropriate accreditation body and must have passed a pair of training courses sanctioned by the Big 3 and offered under the auspices of the Automotive Industry Action Group (AIAG).

Finally, people conducting audits should be (as should all other employees) "qualified on the basis of appropriate education, training, and/or experience, as required."

TECHNICAL REQUIREMENTS

1. Quality system auditors must be "independent of those having direct responsibility for the activity being audited."
2. Registration auditors must:
   - Be recognized and qualified as described in accreditation body criteria.
   - Undergo and pass sector-specific training in QS-9000 and *Quality System Assessment*, offered by the Automotive Industry Action Group (AIAG). The training must be confirmed by a certificate.
   - Have relevant industry experience as defined by the accreditation body.

TECHNICAL GUIDELINES

ISO publishes a guidance document that recommends a set of qualifications and training for quality system auditors: ISO 10011-2 (1991): *Qualification Criteria for Quality System Auditors*. This document itemizes the ideal attributes for quality system auditors. They include:

- Education—Completion of at least secondary school.
- Training—To "the extent necessary to ensure competence in the skills required for carrying out audits." This includes knowledge of:
  - The Standard.
  - Auditing techniques.
  - Audit management techniques.
- Experience—Auditors should have a minimum of four years' full-time appropriate practical workplace experience, at least two years of which should have been in quality assurance.
- Personal attributes—Fairness, persistence, objectivity, unflappability, and reasonableness are essential.

When applied to full-time professional quality system auditors (such as people employed by registration bodies), these guidelines are augmented by additional guidelines published by bodies that provide auditor accreditation (the Institute for Quality Assurance in the United Kingdom; the Registrar Accreditation Board in the United States).

95. What is the typical QS-9000 registration audit like?

CAPSULE ANSWER

Assessors do a thorough, fair, and objective job of assessing a system to evaluate its compliance with QS-9000 requirements.

How about stressful? Nerve-wracking? White-knuckle time? The audit can be all of these. Even if a company is very well prepared—or has been through many other kinds of audits—a QS-9000 registration audit can be one long Maalox moment.

First, let us differentiate between a QS-9000 registration audit and the typical "supplier quality assurance" audit to which many QS-9000 candidates have been exposed in times gone by. In the typical supplier quality assurance audit, the auditors spent the majority of their time in the conference room. They read reports, looked at charts, and talked to the quality manager, quality engineer, and other management and quality personnel. QS-9000 auditors spend most of their time out on the floor. They want to see what is *really* going on, not just what is claimed on paper.

THE PRELIMINARIES

Usually, before coming to a site, the QS-9000 auditors do a "desktop study" of the company's quality system. They review the *Quality Manual* (Level 1) and, sometimes, the Level 2 documents (SOPs). They provide a report of their findings. If there are compliance problems, the company, as a rule, is required to correct them by the time of the registration audit.

Also, in advance, the audit team provides a schedule and plan for the registration audit. The plan is not etched in stone. Findings as the audit progresses can influence the schedule. But it offers a tentative outline for what they are going to cover, and when.

When the auditors arrive at the site, they begin with a brief meeting with management. They discuss the audit schedule and scope, and they answer any questions that may arise.

GATHERING DATA

The audit team usually starts with a quick walking tour of the site, to get oriented. Then, in accordance with a schedule, they fan out across the site, escorted by the designated "guides." These guides should be members of the internal audit team; it is good experience for them. But it is not the guides' job to answer questions or direct the audit team. The guides' role is to get the assessor access to departments, people, and records.

Assessors visit virtually every department of a company. They work off a checklist based on the *Quality Systems Assessment*. Their job is to assess the effective implementation of the QS-9000 system and assure themselves that there are no noncompliances with the QS-9000 requirements. No area, no employee, no function of the company is off limits.

Registration assessors:

■ *Question employees.* They ask about the quality policy, request explanations of their duties, and inquire about their training.
■ *Observe activities.* Always, they are comparing what they see with what the quality system says about the quality practices. Are they real? Are they in place? Are they being followed consistently? Are they effective?
■ *Check records.* This is the most objective form of evidence in support of findings of compliance as well as noncompliance.

They do all these activities on a sampling basis. There is no way, even with the amount of time allotted, to cover any particular activity with complete depth. Instead, assessors "sample" across activities, employees, and records. After they have satisfied themselves that the area in question is in compliance, they document evidence of that fact and move on promptly. They do not engage in fishing expeditions.

If they find, or suspect, that there is a noncompliance, they pursue the matter until they have gathered sufficient evidence one way or the other. And they will almost always alert the escort to the possibility of a noncompliance.

How should a company deal with an assessor? Conventional wisdom says that nothing should be volunteered. "Just answer the questions." If a company chooses to be that way, it may create an adversarial atmosphere where there really should not be one. It is anticipating that the assessor is "out to get us." Most assessors are not. (A tiny handful—maybe. But usually the registrars promote these characters into management sooner or later.) Let the system be fairly and thoroughly audited. The smart route, in the long run, is to be as open and honest with the assessor as possible.

Should escorts or others argue with the assessor? By all means. Assessors will often ask devil's-advocate questions that challenge the many judgment calls that were inevitably made while implementing the system. The worst mistake would be to back off and meekly accept an implied criticism or challenge to the system. Stand up and defend it!

If a company has done its homework, understands the requirements of the Standard, and has made a good-faith effort to comply, the assessor has no choice but to take into account the unique workings of the business and the situation in making his or her judgment. This does not mean that an assessor has the power to waive the requirements. But the assessor is supposed to do a fair, balanced, and objective job of interpreting the requirements in light of a company's characteristics. The assessor can only do that when the company's decisions are explained. If the assessor has the company cold, the internal audit team can only swallow hard, own up, and move on to the next thing.

What the assessor will not do—must not do—is give advice or direction about the system. That is consultancy, and consultancy is forbidden. Assessors are limited to evaluating a system and its compliance with the requirements. They are not allowed to tell how to fix what may be wrong.

Periodically, during the audit, the assessment team will meet privately to review their findings and revise their schedule. They may revisit an area or function once or twice. Or, they may schedule interim meetings to brief management of their findings. However, most of that contact is reserved for the "final phase" of the assessment.

THE GOOD NEWS, THE BAD NEWS

When the data-gathering phase has ended, the assessment team meets to compile their findings. They strive toward positive as well as negative findings. But the negative findings are more important, and they do a thorough job of identifying these and documenting them with objective evidence.

Most registrars rank noncompliances as major or minor (Question 96). They are also required to present "opportunities for improvement," sometimes called "observations."

At the closing meeting, the assessment team presents their findings. Noncompliances are documented on noncompliance reports, and virtually every company will have some noncompliances—as few as a half-dozen or as many as 80! However many there are, they must be closed out before registration can be conferred.

Normally, major noncompliances require a revisit by an assessor for close-out. Minor noncompliances can be closed out without an on-site reaudit, as long as the company provides adequate objective evidence that is then confirmed at the first surveillance assessment. (Practices among registrars vary somewhat here.)

A company cannot consider itself registered—or proclaim itself to be—until all noncompliances from the registration audit have been closed out.

How does the company know how well it did? The fewer majors, the better. No majors at all can be considered a real victory. But keep in mind that there is no real way to "fail." Not if a system has been effectively implemented.

96. What happens if the auditors find noncompliances in a company's quality system?

If is the wrong word. The auditors *will* find noncompliances (sometimes called "nonconformities"). They always do. It is not the end of the world. Unless a company gets multiple major noncompliances (see below), it is not going to "fail." In most cases, the worst thing that can happen is that registration will be delayed until whatever problems were found have been fixed.

> **CAPSULE ANSWER**
>
> Noncompliances are a fact of life. Because they represent an opportunity to improve, they should be viewed positively—even if, as sometimes happens, the appearance of several "major" noncompliances delays registration for a time.

QS-9000 assessors are trained to do a very thorough job of documenting their findings with objective evidence. Before they reach a conclusion, they gather as many facts as they can. To do this, they mention the *possibility* of noncompliances whenever they find them during the audit. This is to give the company and its people a chance to respond and explain.

Do not hesitate to explain and to defend the system whenever it is challenged. Some assessors challenge something just to play a devil's advocate role. More often, they need additional information. They want to render a judgment that is fair and is supported by evidence, and they can only get the facts by asking for them. The worst thing anyone can do, when an auditor challenges something during an audit, is to bow his or her head, swallow hard, and meekly defer. (Actually, the very worst thing anyone can do is say, "Oops, I guess we sort of forgot to meet that requirement there. Sorry.")

Another thing to consider is the fact that every company is different. The Standard, prescriptive though it is, has a lot of "wiggle room." There is a certain amount of latitude for compliance in many areas. In implementing a system, a company make dozens of "judgment calls." Count on assessors to probe those judgment calls and judge for themselves how well the system meets the intent of the requirements. A company's own people need to understand how the QS-9000 system works in their particular area. There is no substitute for this!

After gathering their data, the assessors retire for a bit, to put together their findings. They write some sort of noncompliances report (the nomenclature varies among registrars) and prepare a verbal briefing of their findings.

Then, at the *closing meeting*, the audit team presents a verbal report of its findings. (They follow this up with a written report later.) The report includes a list of noncompliances, supported by documentary evidence. If the team is doing its job, there will be no surprises at this meeting. They will have pointed out potential noncompliances to their escorts during the audit itself. They do the verbal report mainly to confirm that they understand what they are seeing. Even if the auditee implements an immediate corrective action, the observation is still noted for the report.

The worst possible outcome is a "fail," which is earned when there is more than one major noncompliance (see below). Other than that, the usual outcome can be paraphrased as "Here's what you need to do to finish the registration requirements." The team will provide a list of noncompliances—major and minor—and agree on the time frame allowed for fixing them (usually, 90 days). Some noncompliances (majors, usually) require a return visit by the auditors for close-out. Others can be closed out via written documentation, which is then confirmed at the first surveillance assessment. This sequence is at the discretion of the audit team. They will thoroughly review all pending activities at the closing meeting.

The audit team specifically does *not* give advice about how to fix noncompliances. They will, however, often offer the benefit of their judgment. If the question is posed: "If I do XYZ, will that take care of the noncompliance?" The auditors will usually say Yes or No.

The audit report will include commentary on the perceived strengths and weakness of the system. It will also present a (sometimes sizable) list of "opportunities for improvement." These are not as actionable as noncompliances, but they are not just for show, either. The company is expected to consider them and act on them when it makes sense to do so. The "opportunities" are assessed later, during surveillance audits, as part of the company's compliance with the "continuous improvement" element of the Standard (Question 27).

TECHNICAL REQUIREMENTS

1. There are five generic types of observations a registration auditor may make (the nomenclature may vary among registrars):
   a. *Compliance*—The assessed area meets the requirements.
   b. *Strengths and weaknesses*—Assessors are expected to comment on these.
   c. *Major noncompliance:*
      ■ Absence of compliance to a QS-9000 requirement (i.e., no system in place to address the requirement).
      ■ Total breakdown of a system intended to meet the requirement (i.e., a system exists but is not operative).
      ■ Any noncompliance that would probably result in the shipment of a nonconforming product.
      ■ Any noncompliance that could either reduce the usability of the product or service for its intended purpose, or could result in its failure.
      ■ Any noncompliance that, in the judgment of the assessor, is likely either to result in the failure of the quality system to ensure controlled processes and products, or to materially reduce its ability to do so.
      ■ A number of minor noncompliances against one requirement (constituting a total breakdown of the system).
   d. *Minor noncompliance:*
      ■ A failure in some part of the system related to a QS-9000 requirement (when it does not fit any of the definitions under "major").
      ■ A single observed lapse in one area of the quality system.
   e. *Observation*—There is an opportunity for improvement.
2. The potential outcomes of a registration audit are:
   ■ "Pass"—No major or minor noncompliances are found:
      – Registration is conferred upon review and approval of the registration body.
   ■ "Open"—one major noncompliance and/or one or more minor noncompliances are found:
      – Registration may be conferred if the noncompliances are resolved within an agreed-on time frame (usually, 90 days). This status is subject to on-site verification, at the discretion of the assessors.

- "Open" status may be converted to "Fail" status if the non-compliances are not satisfactorily resolved within the agreed-on time frame.
■ "Fail"—More than one major noncompliance is found:
  - The audit must be reconducted.

TECHNICAL GUIDELINES

The same scheme applies to surveillance assessments. The registration body will issue the same kinds of findings, and the outcome possibilities (pass, open, or fail) are the same.

97. What must a company do to keep QS-9000 registration?

Two things: First, deal effectively with the outcomes of the semiannual surveillance assessments, and, second, pay the registrar's invoices.

Of these, the invoices are easier to deal with. The surveillance assessments are another matter.

> **CAPSULE ANSWER**
>
> Surviving surveillance assessments takes consistent, persistent work—and not just for the week leading up to the audit.

Every six months or so, the registrar reassesses a part of the QS-9000 system. The company always knows when the reassessments are coming. The schedule is arranged in advance. What is not necessarily known is the part of the system that the registrar plans to reassess. That is announced when the assessors walk in the door. As with registration audits, the entire system is an open book.

The company must resolve, in an agreed-on amount of time, any non-compliances that arise from surveillance assessments, and they must be closed out by the registrar. Otherwise, the registration is put in jeopardy.

What factors make surveillance assessments such a challenge?

FACTOR 1: THE INFAMOUS "LETDOWN" SYNDROME

Once a company has become registered, a "letdown" syndrome often kicks in. "We made it!" goes the theory. "Now we can ease off." Wrong. The company must keep working its system. That first six months will fly by—and there they will be, the assessors knocking on the door again, poised to reassess part of the QS-9000 system.

If the company has faithfully and conscientiously worked its system, no problem. If it has not, it can be in real trouble. It is a lot easier to work the system day by day than to recover from the trauma of a nasty surveillance assessment.

It is like taking care of a swimming pool. If you do 5 or 10 minutes of work every day, it is easy to keep the water clear and clean. But you have to do that 5 or 10 minutes of work every single day. If you do not, then you get to do 2 or 3 days worth of work to clear your pool of all that lovely lime Jell-O.

FACTOR 2: IT DOESN'T GET EASIER

In some ways, a registration audit is among the easiest reviews a company will ever undergo. Surveillance audits tend to get tougher, pickier, more specific. Why? Registrars want to see that a system and process are improving. That is the whole point of the effort.

Surveillance assessments are one of the three "reinforcement mechanisms" of QS-9000 (Question 13). But they are more than that. They are tangible evidence that a quality system is here to stay, as long as a company wants to stay registered.

98. What should a company do if it doesn't absolutely have to register now?

Breathe a sigh of relief—and then get busy. The fact that QS-9000 is not mandated for the company now does not mean it will never be.

> **CAPSULE ANSWER**
>
> Even if a company is not under the QS-9000 gun now, it should take steps to educate itself and prepare—just in case.

Anyone reading this book is probably connected to a company that is in the QS-9000 arena in some sense. Maybe customers are not requiring registration now, and maybe they never will. Gamblers who are feeling lucky might do nothing and cross their fingers. Forward-thinking and proactive professionals will not want to be caught flat-footed. They will want to be ready.

Here are some ideas.

1. *Implement ISO 9000 now.* There are several good reasons to seri-
ously consider implementing an ISO 9000 system now.

- Unlike QS-9000, ISO 9000 applies to virtually any organization,
making any product or service, anywhere in the world. Even if
QS-9000 does not fit a company, ISO 9000 certainly does.
- QS-9000 is simply ISO 9000 plus a large number of more prescriptive, automotive-industry-related requirements. The implementation, audit, and registration processes are virtually the same.
Consider ISO 9000 as a "logical first step" to QS-9000.
- ISO 9000 is something that can be done now to put the company
well ahead of the game if and when a customer requires implementation of QS-9000. (Customers themselves may not have decided what
they want their suppliers to do. The first step of implementing ISO
9000 may dissuade them from requiring QS-9000—which is all to
the good for the suppliers.)
- Registering to ISO 9000 has the additional and quite desirable effect of helping to improve the company's own performance!

2. *Find out what others in the automotive field are doing.* Quality system
registration (whether to QS-9000 or ISO 9000) tends to impact specific industries all at once. Trade magazines suddenly start talking it up. Papers on
it are presented at trade shows. What is the "buzz" in the company's particular niche of the industry? Is ISO 9000 or QS-9000 being discussed? What
are people saying about either one? Is there a chance that a major competitor might get registered and thereby increase its competitive strength?

The answers to these questions will help with a decision on whether
QS-9000 and ISO 9000 should be part of the strategy for the years ahead.

3. *Get educated.* QS-9000 *is* probably as mind-numbingly boring as
watching paint dry in Flint. But if it is becoming a factor in the industry,
there is no such thing as knowing too much about it.

- Read other books on QS-9000 and ISO 9000. There are many good
ones that go into far more detail about their technical aspects than
this book does.
- Buy and read *Quality System Requirements QS-9000*. This is the
bible. No book, not even this one, is a substitute for it.
- Attend classes and seminars. Many companies offer "internal
auditor" and "lead auditor" training courses, as well as courses on
QS-9000 implementation, document writing, and awareness. Of
these, all but the "lead auditor" course are worth considering. The

typical lead auditor course is a very intensive five-day course that (sales claims notwithstanding) is beneficial only to people who plan to make quality systems auditing a regular professional pursuit.

■ Subscribe to Internet discussion groups on ISO 9000 and QS-9000. These are listed in Appendix C.
■ Subscribe to industry publications. A list of these is also in Appendix C.

4. *Benchmark a company that is going through the process.* The best way to learn about QS-9000 is to go through the process. If that is not possible right now, take the time to find a company (not a competitor) that is implementing a system and preparing for registration. Ask to meet, from time to time, with at least the management representative, to find out what the experience is like.

Anyone fortunate enough to be able to arrange this will gain greater knowledge than is available from all the other sources combined.

PART SEVEN

After Registration

99. Once it is safely registered, how can a company best capitalize on the QS-9000 quality system?

CAPSULE ANSWER

Make the QS-9000 system an active, visible part of the business and its image—externally, as well as internally.

In time, a QS-9000 quality system brings its own rewards. This presumes, of course, that the company does the following:

1. Implements the system fully (no shortcuts, halfway measures, cut corners, or other funny business).
2. Operates it in a continuous manner into its second or third year. ("But how do you know, when it's only been around since 1994?" Because of experience with ISO 9000.)

Here are some of the rewards that a company can realize from QS-9000 registration:

- Retain the business of key customers (90 percent of the time, this is the main reason companies get into QS-9000).
- Compete effectively with other companies registered to either QS-9000 or ISO 9000. (This could offer an edge in competing for business based in the European Union.)
- Acquire new business from other customers who require (or prefer) QS-9000 registration.

There are other actions a company can take to capitalize on QS-9000 registration:

- Place press announcements in all publications that serve the targeted marketplace. Stress the benefits of QS-9000 registration and the implications it will have for the company's business worldwide. Make sure to communicate that registration to QS-9000 is also registration to ISO 9000.
- Reproduce the registration logo on business cards, letterhead, brochures, annual reports, and marketing materials.
- Promote the registration status within industry trade groups.
- Prepare briefings for key customer accounts. Even now, despite the determined efforts of various writers and consultants, knowledge of QS-9000 is fairly limited. Explain QS-9000 in some detail; more important, stress to clients the benefits of doing business with a firm whose quality system is registered.
- Make QS-9000 a major component of sales presentation materials. When going after new accounts, rest the quality message on the company's QS-9000 registration.
- Publicize any quality system success stories to all appropriate stakeholders of the business.
- Encourage suppliers to register to the Standard—and not just to carry out "supplier development" as required by QS-9000 (Question 43). Make a full and steady effort to:
  - Audit key (critical) suppliers against the QS-9000 requirements.
  - Offer these suppliers training, briefings, and orientation in the requirements.

The importance of supplier development to a QS-9000 level cannot be overstated. Just as the Big 3 feel it is vital to deal with their key suppliers on the common ground of QS-9000, so it is important for those suppliers and others to deal with their key suppliers in accordance with a commonly understood standard.

As one of the ISO 9000 guidance documents states:

Mutually-advantageous partnership arrangements between purchaser and supplier [can supplement] third-party audits. Such partnerships focus on mutual efforts toward continuous quality improvement, and the use of innovative quality technology. In instances where purchaser/supplier partnerships are fully developed, third-party certification often plays an important early role . . . . [This] may become relatively less important as the partnership develops, and progresses beyond the requirements of the [QS-9000] contractual standards.

100. How can a company tell whether its QS-9000 system is really working?

CAPSULE ANSWER

The QS-9000 system is really working when employees at all levels take it for granted.

When it becomes not just tolerated, but accepted as an automatic, reflexive part of "the way we do things here." The term for this is "transparency." When a QS-9000 system is really working, it is transparent in the organization.

But (surprise!) this does not happen overnight. It usually does not even happen prior to registration. Up to that point, and beyond, the QS-9000 system is new, different, and intrusive. It forces people to change—most, in small ways, but some in large ways. People resent being forced, and they tend to resent all change. Some resist, at least passively. Others ignore the change and hope that it will go away.

They can run, but they cannot hide. Once the internal audit process starts, employees are confronted with the quality system, and their obligations under it, on a regular basis. Most elect to join 'em, having failed to beat 'em. And that is the big turning point.

Once the euphoria of registration has passed, and the company has a couple of successful surveillance assessments under its belt, the QS-9000 process starts to become transparent. Here are some definitive clues that this is happening:

- Virtually no major noncompliances are written during internal quality audits. When the program first starts, internal audits will routinely turn up quantities of "majors." A few may still slip through, even after registration. By the time the company reaches transparency, majors tend to be past history.
- Middle managers are routinely writing Corrective Action Requests. This is a strong sign because the typical middle manager tends to address problems on a solo basis. When middle managers begin working through the system, it is a sign that they (finally) see value and merit in it.
- Management reviews become strategic in nature rather than tactical. The first few management reviews are extremely tactical, do-it-by-rote exercises. Later, after implementation and registration, senior management tends to realize what excellent tools these reviews are not only for reviewing the quality system processes, but also for improving them over the long term.

- Employees watch the numbers. The system tracks performance against defined goals. A wise management echelon communicates the "score" to employees on a regular basis. Transparency is reached when employees keep an eye on the numbers themselves.
- Procedures and other quality system documents are routinely being reviewed and updated as processes change and improvements are put into place. Until transparency, documents tend to lag behind changes—sometimes for a long time—and are not caught except by internal audits. When transparency sets in, people automatically take care of the documents that pertain to their jobs. They want them to be right.

Unfortunately, some companies never reach the transparency stage. This happens to companies that implement "for the certificate only," or "to get the customer off our back." They do the minimum needed to get registered. They allot meager resources to implementation and maintenance. They communicate to their employees, by word and deed, that, lip service notwithstanding, "This is just a big joke and it really doesn't matter."

Their QS-9000 system is then a cost rather than a benefit. And that is a shame.

101. What is the one key thing a company can do to make the QS-9000 system work well?

> **CAPSULE ANSWER**
>
> Set an aggressive quality-related goal and aim the QS-9000 system (and the entire organization) at meeting that goal.

A well-implemented, contributing QS-9000 system is more than just a "certificate on the wall." It is a system for improving company performance.

Ultimately, to survive as a business, a company must meet customers' needs. Its ability to do that drives everything else: whether it makes money, how well it competes, how fast and large it grows.

The QS-9000 system is a blueprint for helping to improve a company's ability to meet customers' needs. It can be optimized—not just today but in the years to come—by doing the following:

1. Identify a specific need that headlines or is high on the list of the company's most important customers. Customers have many needs, but narrow the choice to just one. It could be something like:

- On-time delivery.
- Fast turnaround time for orders.
- Low variability around target values for critical characteristics.
- Absence of visual defects.
- Fast installation.

2. Develop a way to measure the company's performance in meeting this need. For example:
   - On-time delivery can be measured as the days (or hours) elapsed between required delivery time and actual delivery time.
   - Turnaround time can be measured as the number of days (or hours or weeks) elapsed between order receipt and order shipment.
   - Variability can be defined as an overall Cpk value (Question 58).
   - Visual defects can be measured by comparing a product to defined standards, where each has a numerical value.
   - Installation time can be measured as the number of days (or hours) elapsed between delivery and full operation of the product.

3. Using that measurement method, identify the company's performance level now (baseline). Resist the temptation to exclude certain products or transactions, or to impose caveats that complicate the measurement process and cast doubt on its objectivity. In other words, don't cheat!

4. Set a specific target or goal for the company's performance. Make it neither easy nor simply "aggressive." Make it borderline impossible (at least as people perceive it today).

5. Update the company's quality policy statement (Question 81) to include the new target as the "objective for quality."

6. Educate the entire workforce, consistently and persistently, on the customers' need and the goal. Encourage them to think about how their own job performance affects the company's capacity to improve its ability to meet the customers' need. Ask for their help, on an ongoing basis, in improving the company's ability to reach the goal.

7. Take the customers' need and the goal into account during all critical quality system activities:
   - Advance quality planning.
   - Management reviews.
   - Business planning.
   - Analysis of corrective and preventive actions.

8. Using the measurement method, publicize the company's "score," very visibly, at frequent intervals. Track trends also.

When the company reaches its goal—as it will, if the QS-9000 system and company team are aimed at achieving it—set the goal still higher. Or, reevaluate customers' needs and set a brand new goal. Always keep the system reaching for something. The company can always do better, and QS-9000 can be the system for making it happen.

Appendix A

ABS Quality Evaluations, Inc.
16855 Northchase Drive
Houston, TX 77060
Voice 713-873-9400
Fax 713-874-9564

AGA Quality
8501 E. Pleasant Valley Road
Cleveland, OH 44131
Voice 216-524-4990
Fax 216-642-3463

American Quality Assessors
1200 Main Street
Columbia, SC 29201
Voice 803-779-8150
Fax 803-779-8109

AT&T Quality Registrar
650 Liberty Avenue
Union, NJ 07083
Voice 908-851-3058
Fax 908-851-3158

British Standards Institution Quality
    Assurance
8000 Towers Crescent Drive,
    Suite 1350
Vienna, VA 22182
Voice 703-760-7828
Fax 703-761-2770

Bureau Veritas Quality
    International
509 N. Main Street
Jamestown, NY 14701
Voice 716-484-9002
Fax 716-484-9003

Centerior Registration Services
300 Madison Avenue
Toledo, OH 43652
Voice 419-249-5268
Fax 419-249-4126

Davy Registrar Services
One Oliver Plaza
Pittsburgh, PA 15222
Voice 412-566-3086
Fax 412-566-5290

Det Norske Veritas
16340 Park Ten Place, Suite 100
Houston, TX 77084
Voice 713-579-9003
Fax 713-579-1360

Entela, Inc.
3033 Madison Avenue, SE
Grand Rapids, MI 49548
Voice 616-247-0515
Fax 616-247-7527

Intertek Services Corporation
9900 Main Street, Suite 500
Fairfax, VA 22031
Voice 703-476-9000
Fax 703-273-2895

KPMG Quality Registrar
150 John F. Kennedy Parkway
Short Hills, NJ 07078
Voice 800-716-5595
Fax 201-912-6050

Lloyd's Register Quality Assurance
 Ltd.
33–41 Newark Street
Hoboken, NJ 07030
Voice 201-963-1111
Fax 201-963-3299

NSF International
3475 Plymouth Road
Ann Arbor, MI 48113
Voice 313-769-6728
Fax 313-769-0109

OMNEX
P.O. Box 15019
Ann Arbor, MI 48106
Voice 313-480-9940
Fax 313-480-9941

Quality Systems Registrars, Inc.
13873 Park Center Road, Suite 217
Herndon, VA 22071
Voice 703-478-0241
Fax 703-478-0645

SGS International Certification
 Services
301 Route 17N
Rutherford, NJ 07070
Voice 800-747-9047
Fax 201-935-4555

Smithers Quality Assessments, Inc.
425 W. Market Street
Akron, OH 44303
Voice 216-762-4231
Fax 216-762-7447

Steel Related Industries Quality
 System Registrars
2000 Corporate Drive, Suite 450
Wexford, PA 15090
Voice 412-935-2844
Fax 412-935-6825

TUV America
5 Cherry Hill Drive
Danvers, MA 01923
Voice 508-777-7999
Fax 508-777-8414

TUV ESSEN
1032 Elwell Court, No. 222
Palo Alto, CA 94303
Voice 415-961-0521
Fax 415-961-9119

TUV Rheinland of North America,
 Inc.
12 Commerce Road
Newtown, CT 06470
Voice 203-426-0888
Fax 203-270-8883

Underwriters Laboratories, Inc.
1285 Old Walt Whitman Road
Melville, NY 11747
Voice 516-271-6200
Fax 516-271-3356

Appendix B

SAMPLE IMPLEMENTATION/REGISTRATION TIMELINE

Gantt chart — ISO 9000 implementation schedule (▓ = scheduled/shaded period, X = milestone)

Task	Nov	Dec	Jan	Feb	Mar	Apr	May	Jun	Jul	Aug	Sep	Oct
Commitment	▓											
Planning	▓											
Create quality manual		▓										
Write procedures			X ▓	▓	▓	▓						
Documentation training		▓	X									
Awareness training		▓	▓									
Write work instructions						▓	▓					
Internal quality audit training						X ▓						
Orientation training							▓	▓	▓	▓		
D-Day							X ▓					
Work the system							▓	▓	▓	▓	▓	▓
Internal quality audits					▓	▓			▓	▓	▓	▓
Management reviews					▓	▓		▓	▓			▓
Select registration body								X ▓				
Preassessment									X ▓			
Registration assessment												X

Plan/Staff — Documentation — Specialized Training — Crunch Time

Appendix C

American National Standards Institute (ANSI)
11 West 42nd Street, 13th Floor
New York, NY 10036
Voice 212-642-4900
Fax 212-398-0023
http://www.ansi.org

American Society for Quality Control (ASQC)
P.O. Box 3005
611 E. Wisconsin Avenue
Milwaukee, WI 53201
Voice 800-248-1946
Fax 414-272-1734
http://www.asqc.org

Automotive Industry Action Group (AIAG)
26200 Lahser Road, No. 200
Southfield, MI 48034
Voice 810-358-3570
Fax 810-358-3253
http://www.aiag.org

International Organization for Standardization
Case Postale 56, CH-1211
Geneva, Switzerland
Voice 41-22-749-0111
Fax 41-22-733-3430
http://www.iso.ch

Internet Discussion Groups Related to Quality Standards

Address e-mail message to:

majordomo@quality.org

Include ONLY the following text in the BODY of the message:

For ISO 9000:   subscribe iso9000
For QS-9000:   subscribe qs9000
For Total Quality Management: subscribe tqm-d

Appendix D

THE "SMELL TEST"

KANTNER'S
Quick, Three-Step
UNIVERSAL
(and utterly infallible)
SMELL TEST
for quality system documents

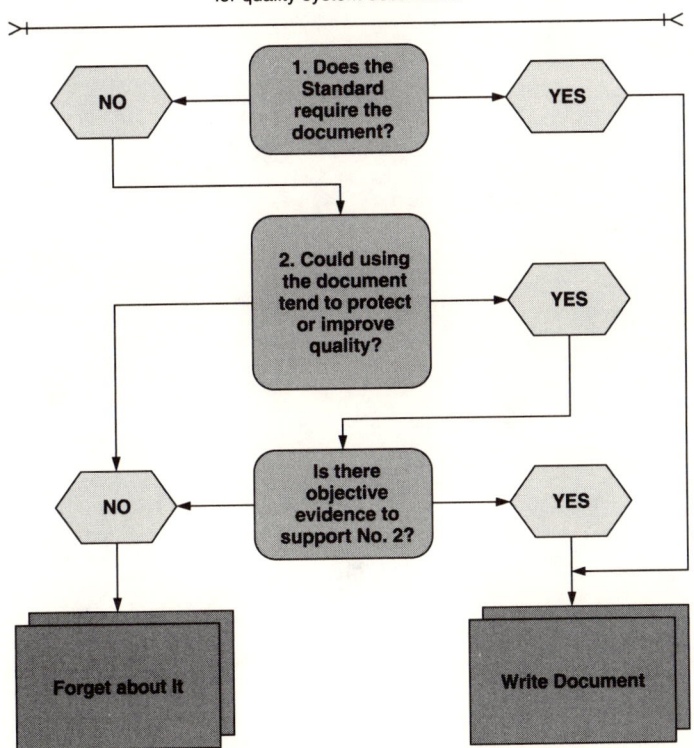

291

Appendix E

SAMPLE PROCEDURE/DEPARTMENT CHART

QS-9000	SOP	Title	All Mgrs	Mgmnt Rep	QC	Sales	Purch
All audits must include these issues in their scopes.		4.1: Quality policy (awareness)	X	X	X	X	X
		4.2.2b Effective implementation of procedures	X	X	X	X	X
		4.17 Work environment	X	X	X	X	X
4.01	01A	Management review	X	X			
4.01	01B	Analysis of company level data	X	X	X		
4.01	01C	Customer satisfaction	X	X	X		
4.01	01D	Business plan	X				
4.02	02A	Quality planning			X	X	X
III.1	02B	Process planning					
4.02	02C	Production part approval process		X	X	X	X
II.1	02D	Continuous improvement	X	X	X	X	X
4.03	03	Contract review			X		
4.04	04	Design control			X	X	
4.05	05	Document and data control	X	X	X	X	X
4.06	06	Purchasing	X				X
4.07	07	Control of customer-supplied product			X		
4.08	08	Product identification and traceability			X		
4.09	09A	Process control					
4.09	09B	Preventive maintenance					
4.09	09C	Safety and environment	X	X	X	X	X
4.10	10	Inspection and testing and inspection and test status			X		
4.11	11	Control of inspection, measuring, test equipment			X		
4.13	13	Control of non-conforming product			X		
4.14	14	Corrective and preventive action	X	X	X	X	X
4.15	15	Handling, storage, etc.			X		
4.16	16	Control of quality records	X	X	X	X	X
4.17	17	Internal quality auditing	X	X	X	X	X
4.18	18	Training	X	X	X	X	X
4.19	19	Servicing			X	X	
4.20	20	Statistical techniques	X	X	X		
III.3	21	Tooling design and management					

Note: There are many different ways to structure and number your Standard Operating Procedures. This is one scheme used by a QS-9000 manufacturing location with design responsibility. By cross-referencing procedures with departments/functions, this chart serves as a training, implementation, and internal auditing tool.

Eng	Tooling	H/R	Prod	Insp	Maint	Whse	Rec	Shpg
X	X	X	X	X	X	X	X	X
X	X	X	X	X	X	X	X	X
X	X	X	X	X	X	X	X	X
X	X			X				
X	X		X	X				
X	X		X	X				
X	X	X	X	X	X	X	X	X
X								
X	X							
X	X	X	X	X	X	X	X	X
	X							
X						X	X	
			X			X	X	X
	X		X		X			
	X		X		X			
X		X	X		X	X	X	X
			X	X			X	X
X			X	X			X	X
			X	X		X	X	X
X	X	X	X	X	X	X	X	X
			X			X	X	X
X	X	X	X	X	X	X	X	X
X	X	X	X	X	X	X	X	X
X	X	X	X	X	X	X	X	X
						X	X	
			X	X				
X	X				X			

Abbreviations used:				
Mgrs	Managers	Purch	Purchase	Maint	Maintenance
Mgmnt	Management	Eng	Engineering	Whse	Warehouse
Rep	Representative	Prod	Production	Rec	Receiving
		Insp	Inspection	Shpg	Shipping

Index